Indianapolis Fire Department

150 Years of Continued Dedication, Honor, and Service to the Citizens of the City of Indianapolis, Indiana

Indianapolis Firefighters Memorial

Indianapolis Fire Department

INDIANAPOLIS FIRE DEPARTMENT

150TH ANNIVERSARY

Pride Never Goes Out of Style

M.T. Publishing Company, Inc.
P.O. Box 6802
Evansville, Indiana 47719-6802
www.mtpublishing.com

Graphic Designer: Elizabeth A. Dennis

Library of Congress
Control Number 2011934982

ISBN: 978-1-934729-66-3

Printed in the United States of America

Cover photo provided by Tod Parker, (phototac.com).

Dedication

Jeff Flick & Lonnie Key

The information for this book was compiled by Indianapolis Firefighters Jeff Flick and Lonnie Key, two proud members of the Brightwood Brigade, for the sole purpose of supplying the most accurate portrayal of the Indianapolis Fire Departments 150 + years of history.

Special thanks go out to all the individuals and organizations that granted us access to their records and photos. This list includes:

Fire Chief Brian Sanford, Howard Stahl, Tim Baughman, Andrew Borton, The IFD Administrative Staff, The Local 416 Union - governing body and staff, Union Office Manager - Lori White, Indianapolis Fire Buffs, Indianapolis Firefighters Retirees Club, Indiana Historical Society, 150 Year History Committee, Christopher Allen (Indianafiretrucks.com), Rodger Birchfield, John Broadwater, Gary Yohler, Francis Shoppenhorst, Roy Dickerson, Joseph Olofson, Fred Hofmeister, Ryan Duffin, John David Fleck, Dale Herr, Carl Hackleman, Mark Rapp, Brett Jackson, Mark Oster, Frank Morgan, Jim Schaffer, Greg Roembke, Scott Olofson, Jim McCormack, Dan Morris, Floyd Sendmeyer, Ed Moore, David Cutshaw, Bob Rumple, Kim House, Bob Lowe, Bob Liming, Joe Krebsbeck, Dr. Jim Brown, Jack Finney, Kris Harrison, Joel Harsin, Jim Major, Floyd Schanz, Eric Scheffel, Bill Sewell, Ted Scott, Rhonda Smith, Jeff Taylor, Rick Van Sant, Frank Wegloski, and Tod Parker (phototac.com).

Additional sources include: City of Indianapolis Records – IFD Headquarters, Cha Cha, Google, *Indianapolis Star*, Local 416 records, Previous IFD yearbooks, and Wikipedia.

Contents

Weyerbacher Apartments

Crystal Flash Fire

Chiefs

Several men and women over the years have stepped-up to lead within the Indianapolis Fire Department. While some firefighters have gained their rank through the merit promotion process, some have been appointed, or handpicked, to lead. There are even positions that just require additional training to obtain including Paramedic, Rescue Technician or Engineer.

Even prior to going to a paid department, someone had to make the big decisions for the rest of the force. In 1843, the designation of Chief of Fire Engineers was given to this position. These are the men who have held this position:

Chief of Fire Engineers

1843-1853 Thomas Smith
1853-1854 Joseph Little
1854-1855 Jacob Fitler
1855-1856 Charles Purcell
1856-1857 Samuel Keeley
1857-1858 Andrew Wallace
1858-1858 John E. Foudry (Resigned 11-16-1858)
1858-1863 Joseph W. Davis (1st Paid Fire Chief)
1863-1867 Charles Richman
1867-1868 George Buchanan
1868-1870 Charles Richman
1870-1873 Daniel Glazier (Only Chief LODD and IFD's 1st LODD, 3-11-1873)
1873-1874 Charles Richman (Only Chief to serve three terms)
1874-1876 Michael F. Fitchey
1876-1878 William O. Sherwood
1878-1882 John G. Pendergast (Only Chief hired with no firefighter experience)
1882-1888 Joseph H. Webster.
1888-1891 Frank L. Daugherty

In 1891, under a new chapter the Office of Fire Engineers was designated as Chief of the Fire Force.

Chief of the Fire Force

1891-1896 Joseph H. Webster (Only Chief to hold both designations)
1896-1901 Thomas F. Barrett
1901-1914 Charles Coots (Longest term, 13 years)
1914-1918 J. Harry Johnson
1918-1922 John C. Louks
1922-1926 John J. O'Brien
1926-1928 Jesse A. Hutsell
1928-1935 Harry E. Voshell
1935-1941 Fred C. Kennedy
1941-1948 Harry H. Fulmer
1948-1952 Roscoe A. McKinney
1952-1956 Joseph F. Hancock
1956-1966 Arnold W. Phillips
1966-1968 John J. O'Leary
1968-1973 David A. Russell
1973-1973 Richard D. Van Sant (Retired 6/28/1973)
1973-1973 John Blazic (Shortest term, 98 days)
1973-1976 Donald E. Lamb (Youngest Chief hired to date at the age of 44)
1976-1979 William J. Patterson
1979-1987 Donald P. Strietelmeier
1987-1992 Joseph D. Kimbrew (1st and Only African American Chief)
1992-2000 Keith D. Smith
2000-2004 Louis A. Dezelan
2004-2008 James L. Greeson
2008-Present Brian L. Sanford (1st Chief from a merged department, Warren Township)

Athletic Club Fire Fatality Funeral

The Last Alarm

In all, 71 firefighters have died as Line of Duty Deaths for the Indianapolis Fire Department.

1. March 11, 1873, Daniel Glazier, Chief Fire Engineer, was appointed chief in 1870. Chief Glazier died when a wall collapsed on him and several other firefighters at the Woodburn-Sarven Wheel Works fire located on Illinois Street, south of South Street. Chief died minutes after being freed from the rubble.

Daniel Glazier

2. November 16, 1886, D. Fremont Taffe, Engine #1, a horse drawn steam engine, – Driver, died from his injuries sustained from a crash at the intersection of North Street and Illinois Street while responding to a fire at a broom factory at 373 North Alabama Street. The engine struck an object in the intersection throwing firefighter Taffe into the street. Firefighter Taffe held onto the reins and was dragged several feet behind the horses before being run over by the apparatus. Firefighter Taffe was carried home where he died the following day from his injuries.

On March 17, 1890, thirteen firefighters lost their lives while fighting the Bowen Merrill Bookstore fire located at 18 West Washington Street. This three alarm building fire started in the basement and grew fast giving firefighters a battle. While working in the structure, the building caved in on itself killing and trapping several firefighters. In all, 10 firefighters died on the scene, two additional died just days after being removed from the rubble, and yet another some two plus years later that succumbed to his injuries. Several additional firefighters were hurt in this fire resulting in them never returning to full duty and taking jobs in the alarm office to complete their careers. The Bowen Merrill Bookstore fire is the deadliest fire for Line of Duty Deaths for the Indianapolis Fire Department.

3. March 17, 1890, Thomas A. Black, Engine #20 – Chauffer

4. March 17, 1890, Andrew O. Cherry, Superintendent of Alarm Telegraph

5. March 17, 1890, George Faulkner, Engine #1 – Chauffer

6. March 17, 1890, Ulysses L. Glazier, Engine #6 – Substitute Firefighter

7. March 17, 1890, George W. Glenn, Reel Company #1 – Pipe man

8. March 17, 1890, Albert Hoffman, Hose Wagon #10 – Private

9. March 17, 1890, David O. R. Lowry, Engine #6 – Private

10. March 17, 1890, Epsy Stormer, Engine #1 – Pipe man

11. March 17, 1890, Anthony Voltz, Truck #2 – Driver

12. March 17, 1890, Henry D. Woodruff, Engine #6 – Pipe man

13. March 18, 1890, John Burkhart, Engine #1 – Substitute Firefighter

14. March 22, 1890, William F. Jones, Engine #6 – Private

15. December 22, 1892, William R. McGinnis, Hose Reel #8 – Captain

Thomas A. Black

Andrew O. Cherry

George Faulkner

Ulysses L. Glazier

George W. Glenn

Albert Hoffman

David O. R. Lowry

Epsy Stormer

Anthony Voltz

Henry D. Woodruff

John Burkhart

William F. Jones

16. August 28, 1893, Benjamin F. Plummer, Hook and Ladder #1, a horse drawn aerial truck – Driver, died from his injuries after being thrown from the truck after hitting a pole. Truck #1, which responded from Station #1, was en route to a box alarm when the accident occurred at the intersection on Pennsylvania Street and Washington Street.

17. December 19, 1899, Webster Robertson, Hose wagon #5 – Private, died from his injuries after hose wagon #5 tipped over throwing Private Robertson while responding to a box alarm. On December 7, 1899, Wagon #5 was en route to 13th Street and Cornell Avenue when it struck a large piece of wood in the street in the 1300 block of Lewis Street.

18. April 19, 1900, Frank E. Mankin, Engine #1, a horse drawn steam engine – Private, died from complications causes from his leg being amputated. While on a fire at the clothing factory located at 13 North Senate Street on March 14, 1900, Private Mankin received severe burns to his leg when a wall collapsed on him and three other firefighters inside the factory trapping them.

19. February 2, 1903, John J. Loftus, Hose Wagon #1 – Private, died when he was run over by a train. While retrieving baled hay for the department horses near West Street and Kentucky Avenue, Private Loftus was thrown from his wagon when the train whistle startled the horses, pulling the wagon suddenly and throwing him from the wagon and into the path of the train.

20. February 4, 1909, Michael Corliss, Hook & Ladder Truck Company #4 – Lieutenant, died from his injuries suffered from being struck by a train. While responding to an alarm at 1800 English Avenue, Truck #4 was struck by a train on State Street just south of Washington Street pushing the truck over 800 feet down the tracks. The three other firefighters on the truck and the horses were uninjured.

Michael Corliss

21. November 14, 1909, W. Russe, Hook & Ladder Truck Company #3 – Private, return from a day filled with fire and smoke exposure that led to pneumonia, resulting in his death.

W. Russe

22. November 8, 1911, Thomas S. Smith, Hose Wagon #16 – Lieutenant, died from his injures sustained from the hose wagon being struck by a street car. Lieutenant Smith, one of IFD's first black firefighters, was en route to an alarm at 21st Street and Northwestern Avenue when the hose wagon was struck in the intersection of 16th Street and College Avenue.

Thomas S. Smith

23. June 19, 1913, Daniel Deveny, Hose Wagon #13 – Private, died after being hit by a train while walking next to the tracks. According to newspaper articles, Private Deveny was on vacation at the time of the accident. Under present day standards, this would not be a line of duty death, however, his name has always appeared on the list.

Daniel Deveny

24. January 5, 1919, Clinton T. Lowes, Hose Wagon #7 – Captain, died from his injures sustained after Hose Wagon #7 and an apparatus from the Underwriters Salvage Corporation collided at New York Street and Delaware Street. Both apparatus were headed to a box alarm at Michigan Street and Illinois Street.

Clinton T. Lowes

25. November 16, 1919, August H. Schafer, Engine #28 – Captain, died from his injuries sustained after Engine #28 was struck by a car at 34th Street and Kenwood Avenue while responding to a residence fire at 3332 North Kenwood Avenue.

August H. Schafer

26. December 21, 1921, George S. Mumford, Truck #5 – Chauffer, died from his injuries sustained after Truck #5 was struck by a car at 14th Street and Meridian Street while responding to a box alarm at 11th Street and Meridian Avenue.

George S. Mumford

27. February 13, 1923, Charles J. Murphy, Engine #6 – Lieutenant, died from complications of a stroke while returning from an alarm run at the English Hotel located on Monument Circle.

28. October 7, 1923, Louis C. Lauth, Squad #1 – Private, died from his injuries sustained after Squad #1 struck a traffic signal pole in the 900 Block of North Meridian Street while responding to a working residence at 2342 North Talbott Street. He was thrown forward into the chemical tank on Squad #1 causing his injuries.

29. January 22, 1924, Fred E. Barlow, Engine #28 – Lieutenant, collapsed and died from a heart attack while fighting a fire at 324 East 36th Street.

Fred E. Barlow

30. March 3, 1924, Wayne Hunter, Squad #2 – Private, died from his injures sustained after Squad #2 and Engine #7 collided at Noble Street and Washington Street. Both apparatus were headed to a box alarm at 37 South Davidson Street.

Wayne Hunter

31. January 18, 1925, David A. Gresh, Engine #14 – Captain, collapsed and died from smoke inhalation while searching for a fire at 2944 North Clifton Street in a vacant building. He had just returned to work after being off for three weeks due to illness.

32. October 19, 1927, George E. Friel, Engine #2 – Private, collapsed and died from smoke inhalation while fighting a residence fire at 2422 Brookside Parkway.

George E. Friel

33. November 18, 1927, William H. Weishaar, Engine #30 – Lieutenant, collapsed and died from smoke inhalation while fighting an apartment fire at 239 North Illinois Street. The fire also claimed the lives of eight civilians.

William H. Weishaar

34. August 3, 1928, Frank M. Yates, Truck #17 – Chauffer, died from his injuries sustained after Truck #17 was struck by a street car at Orange Street and Madison Street while responding to a reported fire at 16 East Caven Street on July 4, 1927. He died over a year later from a cerebral hemorrhage caused by the accident.

Frank M. Yates

35. November 1, 1931, Lewis Stanley, Engine #16 – Lieutenant, died instantly from his injuries sustained after Engine #16 was struck by a car at 16th Street and Central Avenue while responding to a box alarm at 15th Street and Lafayette Street.

Lewis Stanley

36. July 9, 1936, George W. Ream, Engine #12 – Lieutenant, collapsed and died from a heart attack while fighting a grass fire at 10th Street and Ritter Avenue.

George W. Ream

37. December 18, 1936, John Hart, Truck #7 – Private, died from his injuries sustained after Truck #7 was struck by a utility company truck at New York Street and Indiana Avenue while responding to a fire at 518 Indiana Avenue.

John Hart

38. January 4, 1937, Thomas Sheehan, Engine #20 – Private, died from his injuries sustained after being blown down a flight of stairs from an explosion that occurred while fighting a grocery store fire at 4810 East Michigan Street.

Thomas Sheehan

39. January 31, 1937, William Arthur, Engine #8 – Captain, died from exhaustion and exposure that he sustained while assisting the city of Evansville, Indiana with flood relief. He was one of 52 IFD firefighters sent to Evansville to assist.

William Arthur

40. October 29, 1937, Edward J. McGinnis, Engine #13 – Chauffer, died from his injures sustained after Engine #13 and Engine #1 collided at Illinois Street and New York Street. Both apparatus were headed to a box alarm at 226 North Meridian Street.

Edward J. McGinnis

41. August 20, 1939, Matthew Dillane, Truck #19 – Private, died from fume inhalation of chlorine gas while fighting a chemical laboratory fire on August 18, 1939 at the cities sewage treatment plant located on South Belmont Street.

Matthew Dillane

42. December 16, 1939, Clifford C. Woods, Engine #1 – Private, died from his injuries sustained after Engine #1 hit a chuckhole and flipped, pinning him under the engine near the intersection of Beauty Avenue and New York Street while responding to a fire near New York Street and Hansen Avenue.

Clifford C. Woods

43. March 23, 1941, Basil E. Reis, Squad #14 – Substitute Firefighter, died from his injures sustained after Squad #14 was ran off the road by another vehicle on March 13, 1941 at 30th Street and Northwestern Avenue while responding to residence fire at 1150 West 31st Street.

Basil E. Reis

44. July 19, 1941, Leo J. Lindauer, Engine #7 – Private, died from his injuries sustained after Engine #7 collided with a gravel truck at 13th Street and Senate Avenue while responding to an alarm at 14th Street and West Street.

Leo J. Lindauer

45. May 30, 1942, Frederick E. Fries, Chief of Battalion #2, collapsed and died from a heart attack while directing suppression efforts at a 2nd alarm fire at 400 East 9th Street.

Frederick E. Fries

46. January 9, 1944, Reuben G. Von Ohlen, Engine #16 – Private, collapsed and died from a heart attack while fighting a residence fire at 1221 North Oakland Avenue.

Reuben G. Von Ohlen

47. March 22, 1944, Herman N. Lewis, Engine #18 – Private, while detailed to Engine #13, collapsed and died from a heart attack while fighting a building fire at 33 South Meridian Street.

Herman N. Lewis

On December 16, 1944, the department shops sent a mechanic to Station #12, where Engine #20 was quartered, to work on the pumper's motor. The mechanic took the engine and the crew for a test-drive. Chauffer Toombs and Private Adkinson both died from their injuries sustained after Engine #20 collided with a car near 4100 East Michigan Street.

48. December 16, 1944, Hubert S.Toombs, Engine #20 – Chauffer

49. December 21, 1944, Harold B. Adkinson, Engine #20 – Private

Hubert S. Toombs

Harold B. Adkinson

50. January 7, 1945, Randall M. Ball, Truck #5 – Private, died from a heart attack on station the morning after fighting several overnight fires.

Randall M. Ball

51. May 31, 1945, Everett L. Jackson, Engine #12 – Private, collapsed and died from a heart attack while fighting a building fire at Fort Harrison Army Base in Lawrence, Indiana. The fire was a result of riot started by over 2000 jailed soldiers for war crimes.

Everett L. Jackson

52. January 17, 1948, Clarence M. Craig, Engine #16 – Private, died from his injuries sustained after Engine #16 was struck by a car at 49th Street and Capitol Avenue while responding to a car fire at 49th Street and Rookwood Avenue.

Clarence M. Craig

53. November 18, 1948, Benjamin Howard, Truck #18 – Captain, while working in the stations workshop, the gasoline blow torch that Captain Howard was using blew up in his hands on November 9, 1948 causing severe 2nd and 3rd degree burns to his hands. Captain Howard died nine days later.

Benjamin Howard

54. October 24, 1952, Arthur Maar, Engine #29 – Private, died from his injuries sustained after falling off the back of the engine while making a turn at Shelby Street and Pleasant Run Parkway South Drive while responding to a shed fire at 500 South Asbury Street.

Arthur Maar

55. August 16, 1954, Morris E. Princell, Engine #23 – Private, died from his injuries sustained after falling off the back of the engine while putting on his turn out coat at 27th Street and Annette Street while responding to a car fire on August 10, 1954. After remounting the engine and returning to work, he later collapsed, was transported to the hospital, and died six days later due to a head injury.

Morris E. Princell

56. January 19, 1956, Harry Miller, Battalion Chief # 6 aide – Lieutenant, died from a heart attack several days after collapsing at Station #20. Lieutenant Miller had just returned from a run and was filling out the report when he collapsed.

Harry Miller

57. May 6, 1958, James R. Welch, Engine #2 – Private, collapsed and died from smoke inhalation while searching for a victim on a residence fire at 2023 North Hovey Street. The house ended up being unoccupied.

James R. Welch

58. January 5, 1960, Joseph Tuohy, Truck #17 – Private, collapsed and died from a heart attack while fighting a building fire at 1001 South Delaware Street.

Joseph Tuohy

59. November 18, 1961, Charles A. Pierson, Fire Prevention Bureau Inspector, died from his injuries sustained after falling and striking his head on a post at a hotel basement inspection at 123 South Illinois Street on November 13, 1961. He died five days later.

Charles A. Pierson

60. August 17, 1963, Roy Pope Jr., Engine #1 – Lieutenant, collapsed and died from smoke inhalation after becoming separated from his crew and running out of air while fighting a 2nd alarm building fire at 1915 West 18th Street.

Roy Pope Jr.

61. August 14, 1969, Charles R. Ends, Engine #28 – Chauffer, while riding out as the officer, died from his injuries sustained from a second explosion at service station while in the basement fighting the fire 1101 East 38th Street.

Charles R. Ends

On January 13, 1973, while responding to a grass fire at 2000 East Minnesota Street, Engine #3 was struck by a car at Pleasant Run Parkway North Drive and State Street flipping the engine and killing Moore instantly. Poole died the following day.

Raymond F. Moore

Jerry A. Poole

62. January 13, 1973, Raymond F. Moore, Engine #3 – Private

63. January 14, 1973, Jerry A. Poole, Engine #3 – Private

64. June 1, 1977, Thomas M. Donoghue, Engine #4 – Lieutenant, collapsed from a heart attack while fighting a residence fire at 100 South East Street on May 31, 1977. He died the following day.

Thomas M. Donoghue

Chief Glazier's Funeral on the Circle.

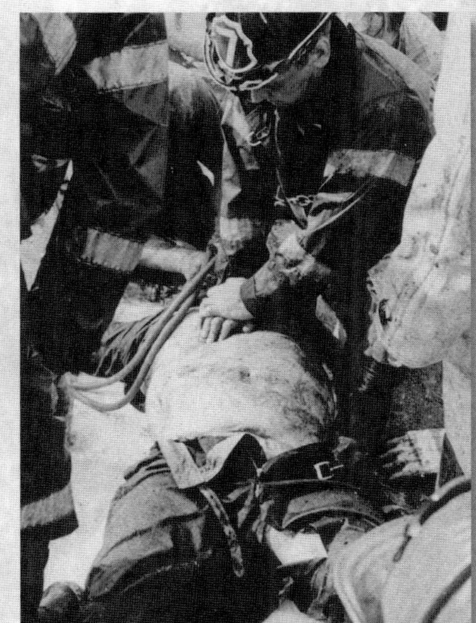

Athletic Club Fire Funeral

65. September 17, 1979, Michael J. Kriech, Squad #29 – Lieutenant, while responding to an automobile accident at Troy Avenue and Madison Avenue. Squad #29 was struck by a train in the 800 block of Troy Avenue on September 10, 1979. He was hospitalized and died a week later.

Michael J. Kriech

On February 5, 1992, while fighting a building fire at the Indianapolis Athletic Club at 350 North Meridian Street, a flashover occurred killing both firefighters. Corporal Gelenius was transported to the hospital where he later died. Private Lorenzano's body was found later and pronounced dead at the scene. Several other firefighters were injured during the flashover.

Elwood M. Gelenius

John J. Lorenzano

66. February 5, 1992, Elwood M. Gelenius, Truck #13 – Corporal

67. February 5, 1992, John J. Lorenzano, Engine #7 – Private

68. May 13, 1994, John Dudgeon, Ladder #14 – Private, died after a lengthy battle from throat cancer from his years of exposure on the job.

John Dudgeon

69. October 26, 1995, John D. Riggins Jr., Ladder #15 – Private, collapsed and died from a heart attack while fighting a residence fire at 1166 Madeira Street.

John D. Riggins Jr.

70. August 13, 2000, Warren J.C. Smith, Engine #14 – Private, died while on a training dive for the department. While performing a training dive, Private Smith got entangled in search ropes. While trying to surface, air clots caused him to go into cardiac arrest.

Warren J.C. Smith

71. June 14, 2002, Paul Jolliff, Squad #10 – Engineer, died while on a training dive for the department. While performing a training dive, Private Joliff became separated from his dive buddy and got entangled in debris causing him to drown.

Paul Jolliff

IFD has lost two firefighters while serving the country in military services.

1. April 8, 1945, Firefighter Charles C. Haas, Engine #27 Private, was killed in action in World War II.

Charles C. Haas

2. August 4, 2008, Captain Gary Henry, Captain of Special Operations, died when a road-side bomb exploded.

Gary Henry

Ladders for Gary Henry

Station #5
122 W. 15th Street

The Settling of Indianapolis Pre 1859

The proud and humble history of the Indianapolis Fire Department (IFD) started quite simply as a way for the citizens of a small city to watch out after each other. From that an organization was born, one that has been a beacon of progress for other departments to follow. The IFD is filled with stories of brave men and women who have committed selfless acts wanting nothing in return. This has been our tradition and way of life for over 150 years. Being a member of the IFD isn't just a job, it's a calling. Being a firefighter isn't something you just do, it's something that you are. If you look at the IFD of 150 years ago and the IFD of today, it would look completely different. We have different uniforms, equipment, apparatus, stations, and responses for example. But with each call to duty, IFD firefighters harness great honor, respect, and dedication as they make sacrifices for the citizens of Indianapolis.

We pay homage to all who have gone before us. We know that we are just a small piece of something bigger and better than we could be on our own. Without looking to build or do something great, the IFD has done just that. As firefighters, all we want to do is help. We show up on the worst days of peoples' lives and do our best to put their lives back together. We go faceless, nameless, and at times thankless, to most that we serve.

Chief's buggy at Station #2.

1679

The 1st department to have paid firefighters was the Boston Fire Department. The first full-time paid department was the Cincinnati Fire Department on April 1, 1853. Indianapolis became the eighth city to have a full-time paid department.

January 17 1825

The 1st reported fire for the city was at Thomas Carter's Tavern, which burned down to the ground.

June 20, 1826

The informal Fire Brigade was formed with John Hawkins as the President and James Ray as the Secretary.

1835

The 1st hand pumper was purchased. The 1st hand pump engine made in America was in 1654 by Joseph Jencks of Lynn, Massachusetts and was offered to the Selectmen of Boston Massachusetts.

Headquarters at Massachusetts Avenue and New York Street.

Ladder #2

We take solace in each other. When all else goes wrong, firefighters still have each other. That's what makes us family. One of many traditions in the fire service is family, a broad but meaningful term. Today firefighters work long hours compared to most jobs. In the beginning, the IFD worked 24 hours a day, 7 days a week and could only go home for meals, family emergencies, sickness or injury. When you spend that kind of time with people, they become like family. Over the years, hours and popular culture has changed but the brotherhood/sisterhood lives on.

Our beginning starts with the building of a city with the Treaty of St. Mary which settled central Indiana. The first white settlers were George Pogue, John McCormick, and their families in 1820. Pogue's cabin was located near East Michigan Street on high ground believed to be on the near west side of Indianapolis. McCormick's cabin was located on the east bank of White River near Washington Street. Additional settlers came soon after and the area came to be known as the Fall Creek Settlement.

The Dalmatian

In the 1700's, Dalmatians were used to protect the horses of stage coach operators. With a growing problem of horse theft, the Dalmatian was used to protect the horses and keep them safe. The Dalmatian was chosen because of its ability to keep up with the stage coaches while in use and the ability to fit under the wheels of the stage coach, making it easier to get from one side of the coach to the other. This allowed the operators to sleep in hotels, instead of with the coaches. Whiling traveling, the Dalmatians would run with the horses warding off other animals that tried to attack the horses.

As the fire service started using horse drawn apparatus, the dogs were used in the same manner. When a horse drawn apparatus arrived on the scene of a fire and firefighters rushed in to put the fires out, it was the Dalmatians that remained outside and stood guard over the horses and apparatus, thus becoming a firehouse staple.

1st Capital Buildings

Vincennes, Indiana

Corydon, Indiana

Indianapolis, Indiana

On May 17, 1820 the Indiana Legislature put together a commission to find a capital city. As the territory became settled, the state wanted the capital city to be more centrally located in the state. The Commissioners visited many places. On June 7, 1820 the Commissioners choose the Fall Creek Settlement to be the location of the state capital. The state had the location of its capital and now all it needed was a name. During the Legislature in January of 1821, Judge Jeremiah Sullivan of Jefferson County said Indianapolis should be the name of the capital. This created quite a laugh, but Samuel Merrill supported it and the name Indianapolis was chosen. Although Indianapolis is the present capital, it hasn't always been the capital. From 1800-1813 Vincennes was the capital as a territory and the first state Capital was Corydon from early 1813-1825.

Indianapolis was given four square miles to form its city. They hired Alexander Ralson to survey it; he also worked on surveying Washington D.C. In his work, he thought four miles was too big and went with one square mile. Mr. Ralson thought the city would be hard to settle because the forest it was cut from was so thick and there were issues with the roads like stumps and roots. When it would rain there would be mud for several days, making huge ruts in the road from the carriages and carts. This made many potential settlers keep traveling. With Mr. Ralson working on the Washington D.C. survey as well, there is no surprise that the two cities resemble one another.

On October 9, 1821 the town lots were auctioned off. Inside the mile square was known as the Corporation and the outside lots were known as the Donation. The auction took place at Nowlands cabin, which was located about where Washington Street and Missouri Street are currently located. The highest price for a lot went for $550.00, which was across the

1835

The Marion Fire Company went into service.

1837

The 1st Station was built for the Marion's.

1840

The estimated Indianapolis population was 2,692 people.

1841

The Independent Relief Company #1 went into service.

1843

Hook and Ladder Truck Company #1 went into service.

1843

The Chief of Fire Engineers position was established.

1849

The Fire Department become the institution of the town. If you wanted to be in politics, viewed as successful in business, or in good public opinion, you needed to be a firefighter.

1849

The Western Libertien Fire Company went into service.

street from the site selected for the courthouse at about Washington Street and Delaware Street on the south west corner. The average price for a lot was $113.00. The out lots sold for either $10.00, $20.00, or $30.00.

The infrastructure of the capital was underway, but fire protection wasn't very high on anybody's priority list until January 17, 1825 when the first recorded fire in Indianapolis took place across the street from the courthouse in a tavern. It wasn't until 17 months later on June 20, 1826 that the city started to make decisions for fire safety. First by passing an ordinance that required all residences of the city to have at least one green fire bucket with their name on it. The city also formed a fire brigade with Caleb Scudder as Captain.

On February 2, 1835 at a meeting of the citizens, a group of loose volunteers formed a more formal fire brigade called the Marion's. Also at this meeting of the citizens, the city decided to pay for half of a hand pump fire engine, 20 buckets, and some new ladders. The state would pay for the other half. A year later in 1836, the state built quarters for the hand pump fire engine. A year after that the Marion's got their first firehouse located on the north side of the then Governor's Circle (now Monument Circle).

Emerald Society

The Emerald Society was formed in the 1800's by professional firefighters of Irish heritage. Most Irishman that traveled to the United States were treated poorly. Becoming a firefighter was a thankless job, but gave the men a place to sleep. With unacceptable work conditions, the men felt it necessary to ban together to protect their own interests forming the Emerald Society. Today the membership is open to all professional firefighters regardless of family lineage, race or gender. The Indianapolis chapter was established in 1994. The mission is to promote goodwill, and camaraderie among the membership, sponsorship of charities, and donations to families of fire fatalities.

In 1839 the downtown canal started to receive water. With more water available for firefighting, the city purchased a new fire engine two years later. The new fire engine cost $600.00. The city paid for half and donations were collected for the other half. When the Marion's got the new engine they divided into two fire brigades. The new fire brigade formed in 1841 was called the Independent Relief Company with John H. Wright as Captain. The Independent Relief Company took the Marion's old engine and named it the "Good Intent." Many of the fire brigades would sing songs while using their hand pumpers allowing them to pump in rhythm. When they needed more water they would sing a faster song to pick up the tempo.

Daily exercising of the horses.

IFD's Age

The Indianapolis Fire Department is older than the following 17 states: Alaska (1/3/1959), Arizona (2/14/1912), Colorado (8/1/1876), Hawaii (8/21/1959), Idaho (7/3/1890), Kansas (1/29/1861), Montana (11/8/1889), Nebraska (3/1/1867), Nevada (10/31/1864), New Mexico (1/6/1912), North Dakota (11/2/1889), Oklahoma (11/16/1907), South Dakota (11/2/1889), Utah (1/4/1896), Washington (11/11/1889), West Virginia (6/20/1863), and Wyoming (7/10/1890).

1850

The estimated Indianapolis population was 8,091 people.

1851

The Marion Fire Company fire house burned down to the ground and was believed to have been started by a firefighter who disliked the living conditions at the station.

The Maltese Cross

In 1113 A.D., The Knights of St. John and the Knights of Crusaders found themselves fighting off the attacks of the Saracens on the island of Malta in the Mediterranean Sea. The Saracens attacked by throwing fire bombs containing flammable liquids at the knights adding a different element to warfare. While fighting off the Saracens and the fires they had started, the knights gave up great sacrifices to hold their ground. Presented by the island of Malta to show gratitude for their victory, the knights flag and an eight-stared cross became the symbol of the knights and became known as the Maltese Cross. Today, firefighters have adopted the cross as a symbol of the sacrifices they make while upholding honor, courage, and dedication to their profession.

Run/Response (R/R) Total

With some sporadic run totals available prior, it was not until 1878 that Chief Pendergast started recording run totals for the department. Over time, the department switched to recording response totals. In addition to response totals, the department started to record total civilian fire deaths in 1947.

Year - R/R - Deaths

1878 – 173	1905 – 1148	1932 – 3323	1959 – 7754 – 13	1986 – 29210 – 18
1879 – 166	1906 – 1203	1933 – 3478	1960 – 8483 – 14	1987 – 34076 – 17
1880 – 188	1907 – 1171	1934 – 3733	1961 – 9131 – 23	1988 – 40792 – 6
1881 – 207	1908 – 1414	1935 – 3159	1962 – 9929 – 30	1989 – 47591 – 13
1882 – 212	1909 – 1323	1936 – 4135	1963 – 12138 – 17	1990 – 51819 – 9
1883 – 214	1910 – 1402	1937 – 2956	1964 – 11599 – 20	1991 – 56197 – 9
1884 – 262	1911 – 1677	1938 – 3235	1965 – 11841 – 33	1992 – 58301 – 13
1885 – 292	1912 – 1777	1939 – 3570	1966 – 13655 – 21	1993 – 57564 – 12
1886 – 347	1913 – 1994	1940 – 4035	1967 – 14360 – 26	1994 – 56394 – 22
1887 – 408	1914 – 2132	1941 – 4063	1968 – 18315 – 30	1995 – 61287 – 12
1888 – 327	1915 – 1991	1942 – 3559	1969 – 17892 – 23	1996 – 94379 [3] – 17
1889 – 302	1916 – 2384	1943 – 4241	1970 – 21734 – 24	1997 – 95635 – 9
1890 – 324	1917 – 2780	1944 – 4188	1971 – 17991 – 26	1998 – 95205 – 14
1891 – 323	1918 – 2724	1945 – 4072	1972 – 15542 – 26	1999 – 92313 – 10
1892 – 435	1919 – 3132	1946 – 4691	1973 – 16459 – 13	2000 – 96021 – 10
1893 – 536	1920 – 3465	1947 – 4670 – 13	1974 – 17436 – 11	2001 – 95260 – 16
1894 – 473	1921 – 3148	1948 – 4854 – 23	1975 – 11428 [1] – 20	2002 – 94736 – 11
1895 – 678	1922 – 4042	1949 – 4458 – 15	1976 – 12300 – 24	2003 – 87748 [4] – 6
1896 – 571	1923 – 4161	1950 – 4295 – 12	1977 – 11037 – 23	2004 – 80790 – 11
1897 – 696	1924 – 3942	1951 – 4884 – 19	1978 – 13686 – 26	2005 – 83977 – 10
1898 – 704	1925 – 4215	1952 – 5763 – 18	1979 – 18713 [2] – 16	2006 – 82602 – 3 [5]
1899 – 957	1926 – 3748	1953 – 6012 – 17	1980 – 21871 – 14	2007 – 121852 [6] – 12
1900 – 1052	1927 – 4127	1954 – 5291 – 25	1981 – 17225 – 16	2008 – 136198 – 16
1901 – 1099	1928 – 4097	1955 – 6408 – 21	1982 – 18530 – 7	2009 – 142829 [7] – 5
1902 – 1098	1929 – 3730	1956 – 6731 – 17	1983 – 24272 – 11	
1903 – 1109	1930 – 4513	1957 – 6728 – 17	1984 – 25544 – 16	
1904 – 1092	1931 – 3602	1958 – 7788 – 28	1985 – 27419 – 17	

[1] Under Chief Lamb, 1st aid runs were dissolved and IFD was an EMS back-up to Wishard Hospital Ambulance Services.

[2] In 1979, IFD started back to providing EMS runs for the city and in 1980 it became mandatory for all firefighters to be trained as EMT's.

[3] In 1996, a significant change was made in the way responses were recorded resulting in a 54% increase in the department's response totals.

[4] In 2003, the department went to a modified response for all alarms runs due to construction of the "hyperfix 65/70" by Indot and the Federal Highway Administration. The modified responses still are in effect today.

[5] Lowest fire related deaths for the city since starting to record them.

[6] In 2007 the department mergered with both the Washington Township Fire Department and the Warren Township Fire Department resulting in a 48% increase in department response totals.

[7] In 2009 the department mergered with the Perry Township Fire Department resulting in a 5% increase in department response totals.

Lost Traditions

1. Station hose towers – Newer advancements in hose manufacturing has eliminated the need to dry-out hoses, which occurred to avoid rot and mildew.
2. Teletype – New dispatch is paperless, leaving no need for teletype.
3. Shining the pole – Lack of use and the fact that a lot of stations do not have them anymore.
4. Making your bed before 10 a.m. – It was a general order and no longer is enforced.
5. Ringing the bells three times when the chief comes by the station to visit.
6. Calling ahead to a station to inquire about putting in for an open spot and respecting the wishes of that station.
7. Water fights.
8. Night watch – Still done at some houses. With the firehouse alerting system, firefighters no longer have to sit next to a radio and wait to hear a run come in for the station.
9. Blood tubs – A reserve of cash set-up by firefighters. It was set-up to help firefighters get cash quick without going to a bank for emergencies.
10. Spanner belts.
11. Life nets.
12. Bowling balls.
13. Staffing board/management board. Now done on computer.
14. Tarps and shutes – Mainly use Visqueen.
15. Gadfly

With the city settling quickly and growing in size, more fire brigades developed. The Hook and Ladder Company organized in 1843. The Western Liberties organized in November of 1849 and used the "Good Intent." The Invincible's were next, forming in 1854. This brigade was made up of mostly Germans and there pumper was called the "Victory." The other fire brigades referred to them as the "wooden shoes." There was also the Union Company formed in 1855; Spirit of 76 formed in April of 1856; the Rovers in March of 1858; and the Young America Hook and Ladder Company formed in May of 1858. Interest in the fire service was so high in 1849 that a developing fire brigade was organized called the O.K. Bucket Company. This was for the young men of the community that weren't quite old enough.

The city itself became enthralled by the firemen and they became the toast of the town. There wasn't a party or event that would have

Water Tower #1 at downtown fire.

Tools of the Trade

The first axes were not attached to a handle or strap and were not hafted. The first hafted axes are known from the Mesolithic Period (ca. 6000 BC).

The first fire hose was put in place in the United States in Boston in 1799 and was made of leather.

The first patented fire hydrant was in 1801 by chief engineer of the Philadelphia Water Works Company, Frederick Graff, Sr. The patent office burned in 1836 making this first a debatable fact.

The first fire helmet, known as a smoke helmet, was invented by Charles Anthony Deane in 1823. In 1836, NYC firefighter, Henry Graticap, introduced the Graticap leather fire helmet that provided significant protection from falling debris.

The first chain saw was invented in 1830 by German orthopedist Bernard Heine.

The first sprinkler system was patented in 1872 by Philip W. Pratt even though Henry S. Parmalee is credited with multiple improvements.

The first station fire pole was placed in service in 1878 in the Chicago Fire Department and was made out of Georgia pine. IFD placed their first pole the same year in Station #2. The Boston Fire Department placed the first brass pole into service two years later in 1880.

The first mention of tarpaulins "tarps" being used for salvage was 1890.

The first pike pole patent, used for firefighting fires, was by Frederick W. Johnson on June 9, 1930. The pole at that time was used to pull down walls to prevent the spread of fires.

The Halligan bar was designed by FDNY Deputy Chief Hugh Halligan in 1948. The Halligan bar is paired with the flat-head axe to form the forcible entry tool called "the irons".

The thermal imagining camera (TIC) technology was introduced into the fire service wide-spread after the 1991 Gulf War once the machines were affordable to produce. The technology was developed and introduced to the military in 1979 by raytheon.

been considered a success without the "fire boys" there. It didn't take long for this to go to their heads. Soon they were getting involved in politics, and unruly behavior, along with other shenanigans like wash-outs. This is when they would go to a "house of ill repute (whore house)" and spray down the inside with their fire hoses. The mayors and council members were becoming upset with firemen for flexing their muscles on many issues. In 1843 the office of Chief Fire Engineer was developed as a buffer between the mayor and all the different fire company's. And in 1843 some of the fire brigades purchased badges. This would be a move to signify power, importance, significance, and unity. It was a symbol of pride. And all the citizens respected them.

In 1849-50 Insurance men put a challenge out to the firemen; one they would come to regret. The insurance men were to give out a silver bugle for the fire brigade that got first water on the most fires and a silver pitcher for the second most first water fires. This would also create a phrase still used today "first due", meaning the first to get water on the fire gets paid. This would cause a string of fires that would involve most of the vacant structures in the city. When a fire company would get first water, they would brag to the other companies and it wouldn't be too long until that company would have a fire that they would be there in no time and so on. This is where the rivalry of "first water" was born. This did nothing to improve relations between firemen and politicians.

There was a growing rift between the firemen and politicians and it was prevalent in the Marion's firehouse. The firemen felt like their firehouse was more for the city and not for them. This did not make the firemen happy and they felt like the living conditions of the firehouse were not tenable and the city didn't care. It didn't really come as surprise that on a warm summer night in 1851, the firehouse burned down. Pretty much all of the belongings of the firemen were able to be saved, but all the city's records were destroyed. No one was ever accused of setting the fire but many believed it to be a fireman. The arson division wouldn't be formed for another 117 years.

1852

The 1st 10 fire cisterns were built in the city.

1852

The Invincibles Fire Company went into service.

1856

The Union Company #5 went into service.

The Fire Association developed in 1856 and was put in place to distribute monies to the different fire brigades. The Fire Association carried about 400 voting members. This also turned into a political machine. The mayor and council members were becoming very tired of the brash firemen. The final straw came at the Fourth of July parade where the mayor fell in love with how uniform all the different fire brigades looked. It did take some time, but a Chief Fire Engineer that wanted a paid fire department was elected. This was not without notice of the volunteers as they cried foul play, but to no avail.

1858

Chief Joseph W. Davis became the 1st paid chief.

1858

The North-Western Fire Company went into service.

Uniforms

The history of the uniform started with just a leather helmet that firefighters had to wear to tell them apart on the fire ground and to protect them from falling debris. Then in 1873 the department recommended the purchase of badges. A year later the department would get its first uniform. It was a red double breasted shirt with dark blue pants. This allowed firefighters to recognize other firefighters, as well as the public, on the fire ground. In 1879 the department purchased a more formal uniform, which is referred to as a class "A" uniform. At the time it was called the "Chief Fitchery" because he pushed hard for these uniforms. It was a single breasted sack coat, indigo-blue in color, with the same color pantaloons or pants, a vest, a blue shirt with a white neck tie or bow tie, and a beaver hat. The chiefs and believed captains wore white shirts and blue neck ties or bow ties. In 1888 the department made the Maltese Cross the official badge emblem until 1943 when they would move the Maltese Cross to the hat badge and design a new more modern badge for the shirts or jackets. In the mid-twenties the department would change the privates class "A" to a white shirt and blue neck tie or bow tie. In 1929 fire prevention changed their badges and coats and members would get service strips for every five years of service which are worn on the cuff of the class "A" uniform.

Then in 1943 the department started giving all members a clothing allowance of $60 in two $30 payments. Five years later the department would make the news with its new uniform policy. This was a time the men would refer to as "all we did was change cloths." All members of the department were required to wear their class "A" to and from work, then change into their "salt n peppers", which looks just like it sounds. They would be in this uniform from 8 a.m.-10 a.m. Then into blue serge uniform from 10 a.m.-6 p.m., Then back to "salt n peppers". One benefit to wearing class "A" uniform to and from work was that members could ride the trolley and train for free.

There were some small changes over the next twenty years like in 1959 chauffeurs got their own insignia patch, which was a steering wheel with wings on both sides. In the 60's the department would do away with "salt n peppers", blue serge, and made changes to the class "A". Instead of the blue serge, the department experimented with uniform rentals. This was thought to be the way to go with the department adding a third shift in 1966, but would soon learn that the quality and desired style wasn't what they wanted which upset the firefighters.

By the end of the 60's the department would settle on two classes on uniforms. The light blue class "B" shirts and dark blue pants, which is comically referred to as "bus driver" uniform and change the class "A" to a double breasted blouse. This is still the uniform worn today. A third class was added, class "C" uniform, which is shorts, sweatpants, tee shirts, and sweatshirts. It started to filter in around the 70's and became more prevalent in the 80's. In the mid 90's the department would go to a fire retardant material in the wake of the athletic club fire. In 2004 the department would go back to 100% cotton clothing and start to let stations design their own station wear with department approval.

Western Station #6 at Washington Street and West Street.

The Foundation 1859 -1909

In August of 1859 the council passed a resolution that would bring a paid fire department. This outraged the volunteers and many fire brigades dissolved. The fire brigades would not give up their apparatus and equipment. At one point the city went two months with no fire protection. The official start of the Indianapolis Fire Department came on November 14, 1859. This made Indianapolis the eighth city to have a fully paid fire department in the United States. The IFD had a limited budget and firemen worked 24 hours a day, 7 days a week. The total number of firemen in the city went from about 600 volunteers to 31 paid firemen. By the middle of 1860 cooler heads would prevail and most of the volunteers gave up their firehouses, apparatus, and equipment.

Many of the volunteers couldn't stay away, when there was a fire they would show up and help the under staffed paid firemen. Others saw the paid department as an opportunity for more city jobs, or at least steady work. The Chief Fire Engineer fell into this category. It had a meager salary of $300.00 a year, but was considered a stepping stone position for one with substantially more money and more political clout.

Now that a paid fire department had been established, it was time for the business at hand of serving. One big change occurred as the nation was divided by the Civil War. Indianapolis firemen were issued side arms because of the threat of Confederate forces burning the city. What is now the state fairground (then Camp Morton), was used as a jail housing Confederate soldiers. It was believed that a famous confederate raider John Morgan was within 3 miles of the city at one point. Firemen stationed

14 November 1859

The Indianapolis Fire Department was started.

1860

Hand Engine Company #1 went into service with four horses, Hand Engine Company #2 went into service with three horses, and Hand Engine Company #3 went into service with five horses.

1860

The estimated Indianapolis population was 18,611 people.

1860

The Indianapolis Steam Fire Department was established on the principle of regular and specified compensation to all employed under the auspices of the department.

1862

A steam engine burned a ton or more of coal per hour while in operation.

Alarm Box 45

Placed at the corner of Meridian Street and Washington Street, Box 45 has been struck multiple times leading to some of the Departments most historical fires including the Bowen-Merrill and Grant Fire.

South Meridian Street Fire

1863
Every member of Indianapolis Fire Department was issued a revolver.

1863
The 1st Central Alarm bell was placed went into service.

1864
The mayor issued 358 certificates for $50.00 to the volunteers for maintenance, protection, and support of families of hardship during the civil war.

1864
The Watchman Tower located atop the Glennblock Building at 25 East Washington Street was placed into service.

1865
The Fire Department office was moved to the 2nd floor of the Glennblock Building at 25 East Washington Street.

in the watchtower had to keep a special eye on the jail to see if any soldiers tried to escape. The thought was that on his way to free the soldiers, he'd burn the city, and kill anyone in his path. None of this ever came about, but there were a tense couple of days when the telegraph lines to the city were cut.

By 1865 the Chief's office was moved into the second floor of the Glennblock Building (located at 25 East Washington Street) which is equivalent to the City County building today. The heads of most, if not all city departments, had their offices in the building. Also, a watchtower was placed in the back in 1864. The city was divided into nine wards and when the watchman would see a fire, he would strike the bell to the cor-

Allen Surgical Institute Fire

Minutes before midnight on January 21, 1892, a fire broke out at the Allen Surgical Institute at the corner of Illinois Street and Georgia Street. The Surgical Institute housed disabled patients as well as injured inmates. Over 300 people were in the building when the fire broke out. In all nineteen people perished and an addition 50 were injured in this fire.

Water Tower #1 at New Jersey Street and South Street.

Capitol Paper Company Fire

responding ward. In some cities colored lens in lanterns where used at night to help with directions to the fire within the ward. For instance, a red lens might mean the north side of the ward, alerting firemen who saw the lantern from the watchtower when their ward bells rung to respond to the north side of that ward. The height of the watchtower needed to be one of the highest buildings in the area so watchmen could see for fires. In some cases watchmen saw other crimes and would report them as well.

What started out with 31 firemen in 1859 had shrunk to 29 by 1870. With a few more runs, firemen were becoming weary of the long hours. In 1870 they were allowed to spend one night at home per week. Up to this point they were only allowed to leave the firehouse for emergencies at home, sickness, injury, and meals. This was very appeasing to the firemen.

1868

The telegraph system was placed into service costing $6,000.00.

1869

The department had 3 steam engines, 8 fire hose reels, 40 firemen and 14 horses.

1870

Firefighters worked 7 days a week and were granted one night a week to go home and sleep.

1870
The estimated Indianapolis population was 48,244 people.

1 June 1871
The Water Company went into service with 200 plugs and 50 boxes.

1871
The new Gamewell system was installed.

15 April 1872
The department held it's first Firemen's Ball.

Marion Fire Company

Station #2

11 March 1873
Chief Daniel Glazier was the 1st Line of Duty Death (LODD).

8 December 1873
Badges were recommended for all firefighters to distinguish them while in public.

1873
Chief Charles Richman became the only chief to serve three terms as chief.

1873
Headquarters was moved to 250 Massachusetts Avenue.

This new time away from the fire house once a week didn't affect their routine at the firehouse. The horses were still exercised everyday by the firemen and did readiness drills 3 times a day, once at 9a.m., noon, and 7p.m. The firemen had to have the horses hitched and ready to go out the door in 3 minutes and would have to stay ready for 5 minutes. The horses got used to this routine and about the time for the readiness drill would stand at the stall door and look at their spot waiting to be put in their place.

The 1870's would prove to be a very pivotal decade for the department. Many changes would come that shaped the department for the future. In June of 1871 the water company put 200 hydrants and 50 boxes into service, which helped tremendously with firefighting operations. A new Gamewell system was put into place, which made receiving alarms much faster and more accurate for both the alarm office and the companies on the run. In 1873, the department ordered new badges and the year after uniforms. The uniforms were very typical of the fire service of the day, red double breasted shirt with dark blue pants.

Fire Gear

Since firefighters have made aggressive interior attacks there has been a need for gear to protect them. It is believed the firefighters started wearing leather hats with brims (cowboy hats) to keep hot water and embers off their cloths and leather gloves to help protect their hands. The heat of the fires led to the need for coats. The members of the IFD in large part in 1874 wore rubber coats that cost $3.75 apiece. Some 11 years later the department would put fire helmets into service to protect members from heat and falling debris. This would be a huge step in the safety of fire ground operations. Over the years firefighters would experiment with different coats, gloves, and helmets. Many firefighters assigned to trucks would wear denim coats with a cotton liner for more mobility when cutting holes in roofs and doing ventilation. This wasn't the best protection when functioning in interior operations. The rubber coats gave more protection and were used much more often which helped to repel water as well.

The IFD used ¾ boots for years for daytime runs. In the evening, they would use night pants, which were very similar to current day bunkers.

In 1976 the department went to Nomex fire coats which gave more heat protection than before. A few years later in 1981 the department went to the Cairns 880 fire helmet, made of plastic, doing away with the metal fire helmets.

In 1993 the department totally went away from ¾ boots when it went to Fire Gear brand. This brought the first tan gear for the department. It cost $1008 per set and two years later each member would receive a second set of fire gear, which was also a first. The IFD would stay with this gear until 2004 when it went with Globe fire gear, which is black in color. This first set of gear would take over a year to get issued with the second set coming in 2005-2006. Part of the hold up with the purchase of the gear and receiving it was due to Globes military contracts being fulfilled before its private sector contracts. This Globe gear cost $1300 a set. By 2009 the total cost to outfit a new recruit with two sets of fire gear and uniforms cost $6211.19.

1873
Hose Reel #1, #2, #3, #4, #5, #6, #7, #8, and #9 went into service.

29 June 1874
An order was made for uniforms to distinguish firefighters from the general population.

1 November 1874
The 1st blue uniform was worn by Chief Fitchery.

21 March 1874
The big fire of 1874 was the first time fire jump across the street starting an additional fire, which was not thought to be possible.

1874
Hook and Ladder Truck Company #2 went into service.

1875
Hose Reel #10 went into service.

1875
Steam Engine #5, #6, and #7 went into service.

19 May 1876
The 1st paid African-American fire company went in service, Hose Company #9, and was staffed by 4 men working 7 days a week manning a reel wagon.

1876
Hose Reel #11 went into service.

26 May 1877
The 1st Rules and Regulations were adopted.

1877
Department staffing consisted of 79 firefighters and 34 horses.

1878
Chief John G. Pendergast became the first chief to start recording run totals for the department.

1878
Chief John G. Pendergast became the only chief hired with no previous firefighting experience.

Station #6 on Washington Street.

1878

The departments 1st fire pole was placed into Station #2. The fire pole was invented in 1878 by Captain David Kenyon of the Chicago Fire Department.

14 June 1879

The 1st General Orders were written for probation offenses.

27 September 1879

Uniforms and beaver hats were ordered for entire staff.

30 October 1879

The playing of cards in the firehouse was prohibited.

Things went on as usual until March 11, 1873 when the IFD was dealt its first big blow. Chief Daniel Glazier became the first Indianapolis firefighter killed in the line of duty. This was not only a huge blow to the fire department, but the city as well. The Glazier family was an Indianapolis firefighting legacy. At the time, Chief Glazier's funeral was the largest the city had ever seen. Mourners stood out to watch the parade go by and to pay respect as businesses covered their windows and doors in black.

In 1874 a large fire downtown was able to communicate to a building across the street which was thought to be impossible due to construction, wide streets, and the ability of the fire department.

The civil war ended in 1863 freeing all African Americans. Even though slavery had ended, there would not be many opportunities for gainful employment for some time. The first African Americans were

First Black Firefighters

On May 19, 1876, Chief William O. Sherwood appointed the first four black men to the Indianapolis Fire Department. The four black firefighters replaced four white firefighters on Hose Company #9, located at 31 West Saint Joseph Street. This became the first all-black fire company and station in the city. Robert Braxton was the foreman with James Graves, Thomas Howard, and Thomas Smith serving in his company. In 1881, Hose Company #9 was moved to 1602 North Carrolton Avenue and was renumbered Station #16 in 1897. In 1922, the company was moved to 441 Indiana Avenue and was renumbered Station #1. It remained as the only black station in the department until full integration on January 1, 1960.

22 November 1879
All fire stations were switched to gas heating.

1879
All stations starting keeping fire records.

1880
A total of 599 fire hydrants were in service.

1880
The estimated Indianapolis population was 75,056 people.

1881
African-American Hose Company #9 was moved to 1602 North Carrolton Avenue.

1881
Chemical Company #1 and Hose Reel #12 went into service.

1 February 1882
All apparatus received gongs.

1883
Chemical Company #2 went into service.

Station 5 at 126 W. 15th Street.

hired in 1876, four to be exact. They worked out of their own firehouse on the west side, Hose Company #9. The IFD would only hire more African Americans if one quit, died, got fired, or retired. Although the IFD was one of the first departments to employee blacks as equal firemen they were not allowed to work at other firehouses.

The department was growing and changing creating the need for more structure. The first written rules and regulations policy was adopted May 26, 1877. We know there were rules before this but not a written policy on how the department would conduct business and themselves. By the end of the decade the IFD would have 56 firemen.

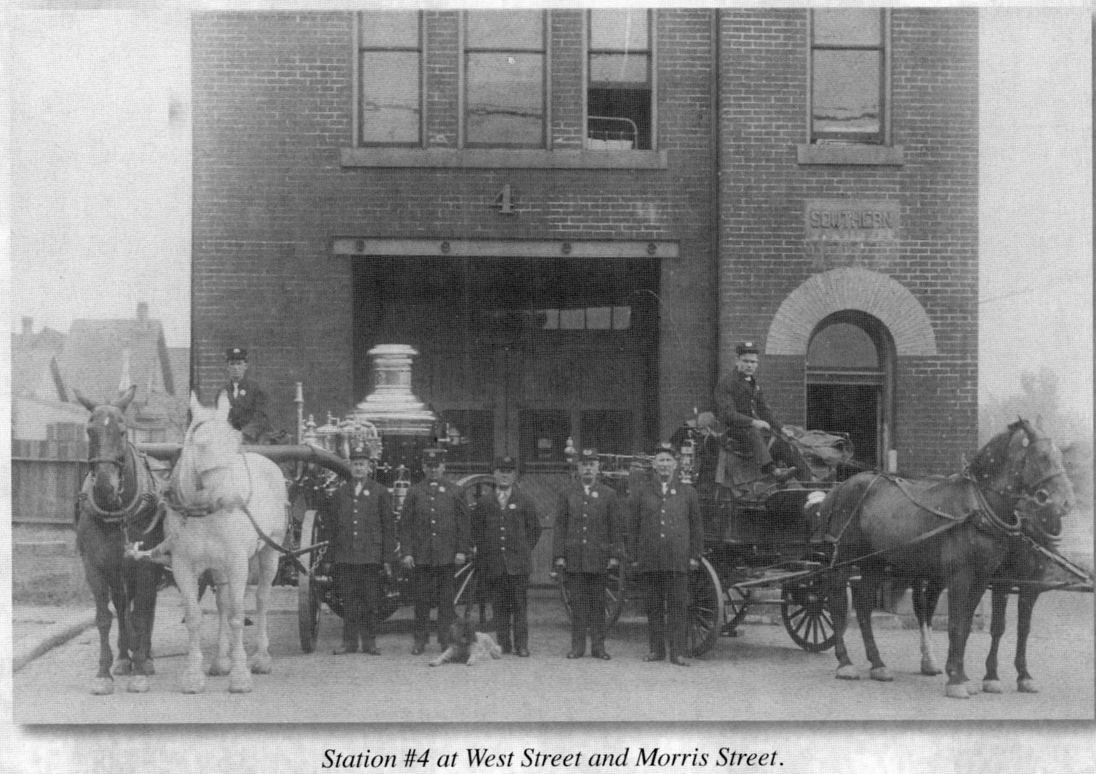

Station #4 at West Street and Morris Street.

1 April 1885
The 1st fire helmets went into service.

1 June 1886
The 1st 24 hours off duty was granted to firefighters.

1888
The wholesale business fire of Meridian Street caused $600,000 in damage.

1890
Department staffing consisted of 89 Firefighters.

1890
Hook and Ladder Truck Company #3 went into service.

1890
Horses in the firehouses had readiness drills 3 times a day, one at 9:00 am, 12:00 pm, and 7:00 pm.

1890
Most horses are purchased at the age of 5-6 years old.

1890
The 1st mention of tarpaulins "tarps" were used for salvage.

1890
The Bowen-Merrill fire killed 13 firefighters resulting in the start of the Board of Public Safety in 1891.

1890
The estimated Indianapolis population was 105,436 people.

6 March 1891
The Department of Public Safety was created.

1891
Chief Joseph H. Webster became the only chief to hold both designations, Chief of Fire Engineers and Chief of Fire Force.

1891
Department staffing consisted of 123 firefighters.

1891
Over 14,000 feet of fire hose was in service not including rubber chemical hose.

1891
The city has the smallest department in the United States compared to other cities it's size.

1891
The Fire Pension Fund was established by State Law as a result of the Bowen-Merrill Fire.

21 January 1892
The Surgical Institute fire claimed 19 lives of patients and inmates.

1892
Chief Webster did away with horse reels pulled by 1 horse, A.K.A. "Horse-Killers"

1892
Hook and Ladder Truck Company #4 went into service.

1893
Chemical Company #3 went into service.

1893
Department staffing consisted of 120 firefighters and 66 horses.

1893
The 1st annual report was released.

1893
The Indianapolis Fire Department covered 21 square miles.

1893
Three smoke protectors were purchased for $300.00 to aid in the rescue of trapped persons.

26 June 1894
The Board of Public Safety granted 7 days of paid vacation for firefighters.

The next ten years didn't produce a whole lot of change in the department other than gongs being added to apparatus to help clear the way on runs (February 1, 1882). The department also put another watchtower in place on the courthouse in June 4, 1883. One change that the men loved was that they finally got an official day off. They got one day off a week, so their schedule went to six days on and one day off. They kept their 24 hour shifts. The rest of the decade went on pretty uneventful.

This would come to a screeching halt on St. Patty's day 1890. The IFD battled a four story building fire at the Bowen Merrill bookstore on West Washington Street. While some of the old volunteers and citizens who thought a paid fire department was a waste of money, heckled and

Bowen-Merrill Fire

On Monday, March 17, 1890, around 3 p.m., a fire broke out in the basement of the Wholesale and Retail Book and Paper House of the Bowen-Merrill Company, located at 18 West Washington Street bringing an alarm from Box 45. While fighting this three alarm fire, the building collapsed in on itself killing 13 firefighters and injuring an additional 18 firefighters. Ten firefighters were pronounced dead on the scene or once removed from the rubble. The 11th firefighter died the following day and the 12th died five days later on March 22. The 13th firefighter died nearly three years later on December 22, 1892 from complications from his injuries.

The following day, Mayor Thomas L. Sullivan issued a proclamation to form the Relief Committee. This committee was established to generate funds for the families that had lost their loved ones in the fire. Members of this committee collected over $52,000 in donation for the Relief Fund. The monies collected were distributed to the families after funeral and healthcare bills were paid. The Bowen-Merrill fire is IFD's deadliest loss of firefighters to date.

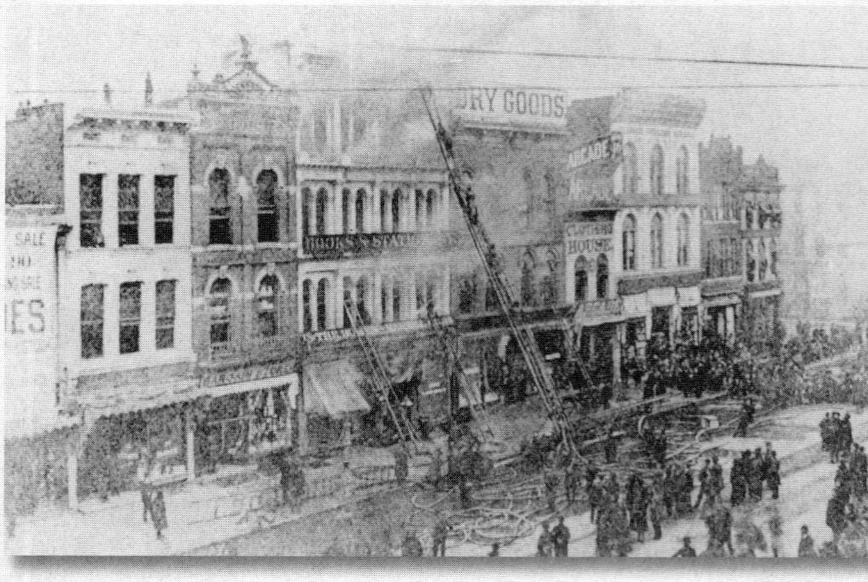

jeered the firemen that were working. Then all of a sudden the building collapsed on itself eventually killing 13 firefighters and wounding another 18. Half of the department was either killed or injured at this fire. Although the public had admiration for firemen, some didn't agree with a paid department, that sentiment would end on this day. The city's hearts broke with the passing of these fine young men and decided to show their respect. City officials, department members, and pillars of the community (such as Eli Lilly) raised funds and passed laws to establish a pension and line of duty death benefits. This came to fruition in 1892. Probably the deepest family affected by this is the Glazier family, who had now lost two family members in the line of duty.

Hose and Chemical Wagon at Headquarters.

1 October 1895

Hose Reel #14 went into service.

30 May 1896

A new water tower placed at Headquarters.

1896

Hose Reel #13 and #17 went into service.

29 April 1897

The West Indianapolis Fire Department was made a paid department when West Indianapolis was annexed by the city of Indianapolis.

1897

African-American Hose Company #9 was renamed Hose Company #16.

1897

Telegraph wiring was replaced with copper lines to all firehouse taking 8 years to complete.

1899

All stations were alerted when an alarm box was activated.

1899

Department staffing consisted of 170 firefighters.

1899

Eight firefighters went before the board for disciplinary action including two for intoxication - one on duty, one for assaulting his Captain - who was fined 15 days of pay, and one being charged with the inability to read and write.

1899

Hook and Ladder Truck Company #5 went into service.

1899

The department had 93 horses, 84 required to move the apparatus and 9 kept as reserves.

1899

The watchtower reported 352 of the 957, or 36.7%, of the departments' runs.

1899

There were 166 cisterns ranging from 700-2000 barrels of capacity, none were connected to the central main.

St. Florian

Florian, a high-ranking Roman officer and by some accounts thought to be in charge of the fire brigade, was ordered to persecute Christians and refused to do so because he secretly was a Christian. Emperor Diocletian sentenced Florian to a "death of faith" – condemning him to be burned alive on a pile of wood. Florian dared them saying "If you do, I will climb to the heavens on flames." The Roman army, scared to test this threat, tortured Florian by flogging and scourging him as he laid half-flayed alive and was eventually thrown in the Ennis River with a stone around his neck to drown.

Since Florian was made a Saint by Pope Lucius III, he has been attributed to saving an entire village from fire by appearing above it and pouring a single bucket of water on the village extinguishing all the flames. St. Florian has also been said to have rescued people from fire through prayer.

1899

There were six department horse deaths, one by collision with a car, one dropped dead in harness, and four by injuries unrelated to firefighting.

1900

The estimated Indianapolis population was 169,164 people.

1 June 1901

Hose Reel #23 went into service.

27 August 1901

The city held the Chiefs Convention for the Midwest.

25 November 1901

Gamewell system was moved the top floor of city hall.

Station #3 at 440 Virgina Avenue.

Finally the department wasn't put on the monitory back burner and the city started hiring more firemen and merging with the West Indianapolis Fire Department in 1897. This wouldn't happen again for almost 110 years. Staffing by the end of the century had climbed to 170 firemen and 93 horses. The communication department in this time frame replaced its telegraph wiring with copper lines, this was expensive and time consuming taking eight years.

Ladder Company #2 at 400 E. South Street.

Station #21 at 1533 N. Roosevelt Avenue.

1901
The 1st electric traffic light was installed in Indianapolis.

1902
Hook and Ladder Truck Company #6 and #7 went into service.

1902
Hose Reel #22 and #24 went into service.

1903
Hook and Ladder Truck Company #9 went into service.

1903
Hose Reel #25 went into service.

1904
Department staffing consisted of 236 firefighters and 112 horses.

1905
Chief Coots recommends prosecution to anyone who maliciously breaks the box alarm glass when no emergency exists.

1905
Fire Station #26 and #27 had a jail cell in the basement which was used by the police department.

1905
The firemen's ball was abruptly halted.

1906
Firefighters were granted 10 days of paid vacation.

1906
Hose Reel #26 went into service.

1906
The city had 289 fire alarm boxes.

27 March 1907
The Board of Public Safety prohibited firefighters from being employed elsewhere while off duty.

14 November 1907
The position of cellar inspector was created to perform building checks.

The new century brought with it a new Gamewell system in 1901 that was put in the top floor of city hall. This was the start of a lot of new technology like the first electric traffic light in 1901. The department appreciated all the modern technology so Chief Coots recommended prosecution of anyone who would maliciously break the box alarm glass. Also, between1900-1904 the rank of Lieutenant was started. The department progressed significantly in this time period. The members were granted 10 vacation days in 1906. The Chief and Board of Public Safety in 1907 felt like the firemen needed to only focus on their job with the IFD and prohibited them from working another job. In 1907 the position of Cellar Inspector was created to do building checks, becoming the first fire prevention position. This time period would see one of the most beloved chiefs, Chief Coots receive the first motorized chiefs' buggy in 1908. By the end of the decade the department had more than 260 members and was growing fast.

Woodburn-Sarven Wheel Works Fire

On March 11, 1873, a fire broke out at the Woodburn-Sarven Wheel Works located at the corner of South Illinois Street and Garden Street. This fire claimed the first Line of Duty Death for the Indianapolis Fire Department when Chief Daniel Glazier was killed. While extinguishing the fire on the second floor, the fire wall collapsed killing Chief Glazier. He is the first chief to die while on duty.

11 March 1908

Chief Coots was the 1st chief assigned to a motorized vehicle.

8 April 1908

Chief Coots orders that "Firemen must be in full uniform, coat buttoned up, blue shirt and white tie and shoes shined."

1908

A total of 2,485 fire hydrants were in service.

1908

Department staffing consisted of 265 firefighters and 121 horses.

1908

The Chief of Engineers salary was $3200.00 a year while other chief's made $2200.00. 1st grade firefighters made $2.75 a day, second grade firefighters made $2.50 a day, and substitute firefighters made $2.00 a day.

1908

The city had 172 fire cisterns with 93 of them being connected to the water main.

Former Station #12 and Station #30.

16 FT.

Wooden Ladders and Iron Men 1910–1959

The progress of the department would continue for the next couple of decades. The department, in keeping with the times, purchased its first front line motorized apparatus Auto Squad 1, which went into service at Delaware Street and New York Street in 1911. The following year firefighters pay went from once a month to twice a month. From here, horse drawn apparatus started to be replaced with motorized apparatus. This started with Engine 28. Station 28 followed by being the first station to be built for motorized apparatus in 1912. Also in this year the firemen received one day off per week. Then in 1914 Chief Coots, still the longest tenured fire chief, retired. The Mayor and Board of Public Safety initially refused his resignation and tried to convince the Chief to stay. The meeting was standing room only and not a lot of dry eyes.

The department continued business as usual, purchasing and putting in place more motorized equipment. By June of 1921 the department was fully motorized and the last horse was retired.

Station #2 on Massachusetts Avenue.

1 April 1910

Auto Squad Wagon #1, the 1st front-line fire apparatus, went into service.

27 June 1910

Hook and Ladder Truck Company #2 closed.

27 June 1910

Hose Reel #28 ran out of Station #26 until 1912.

1910

The estimated Indianapolis population was 233,650 people.

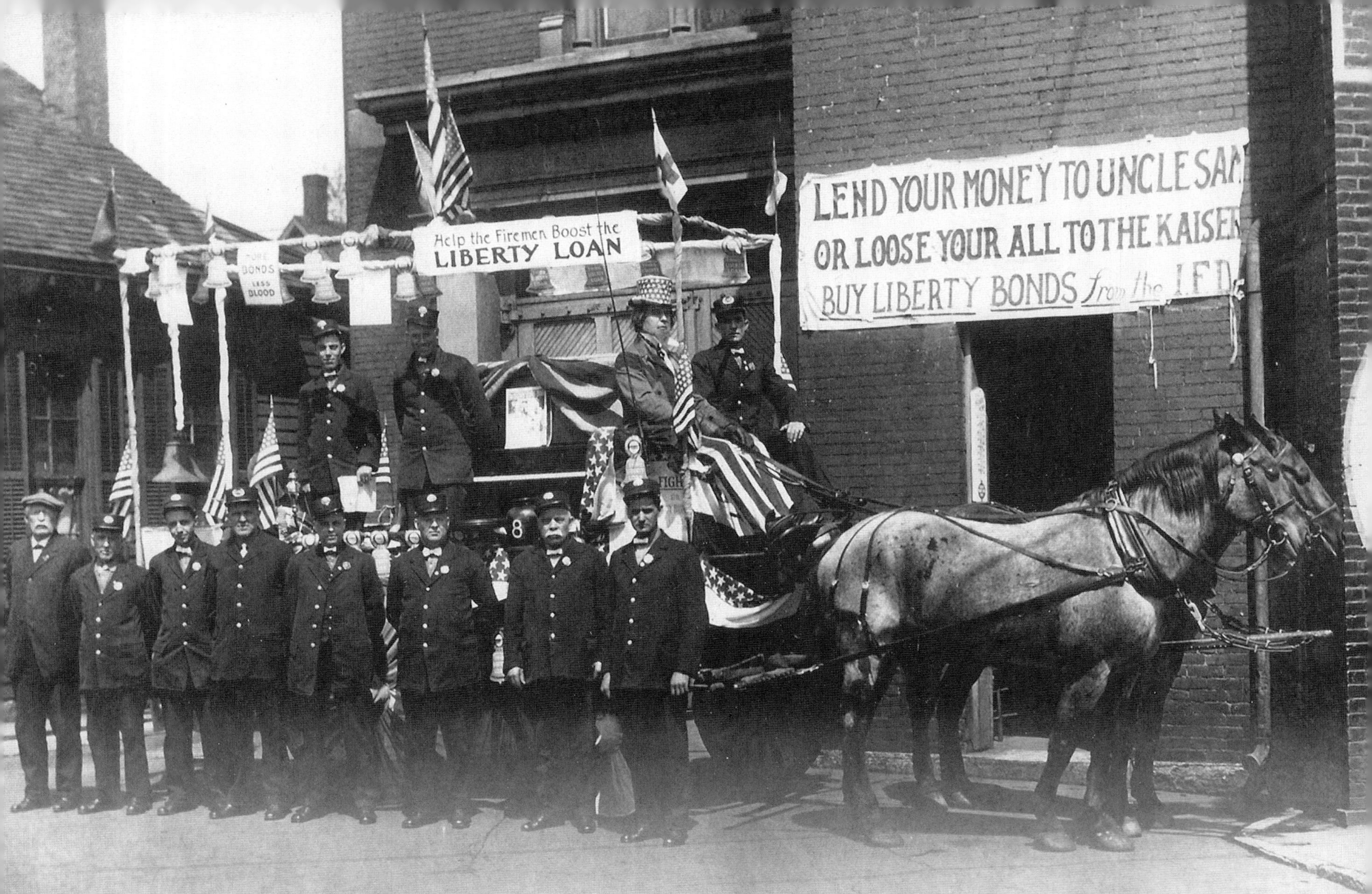

1 October 1911

The 1st Fire Prevention day was observed.

1 June 1912

Pay went from once a month to twice a month.

1912

Engine #5, #7, and #28 went into service.

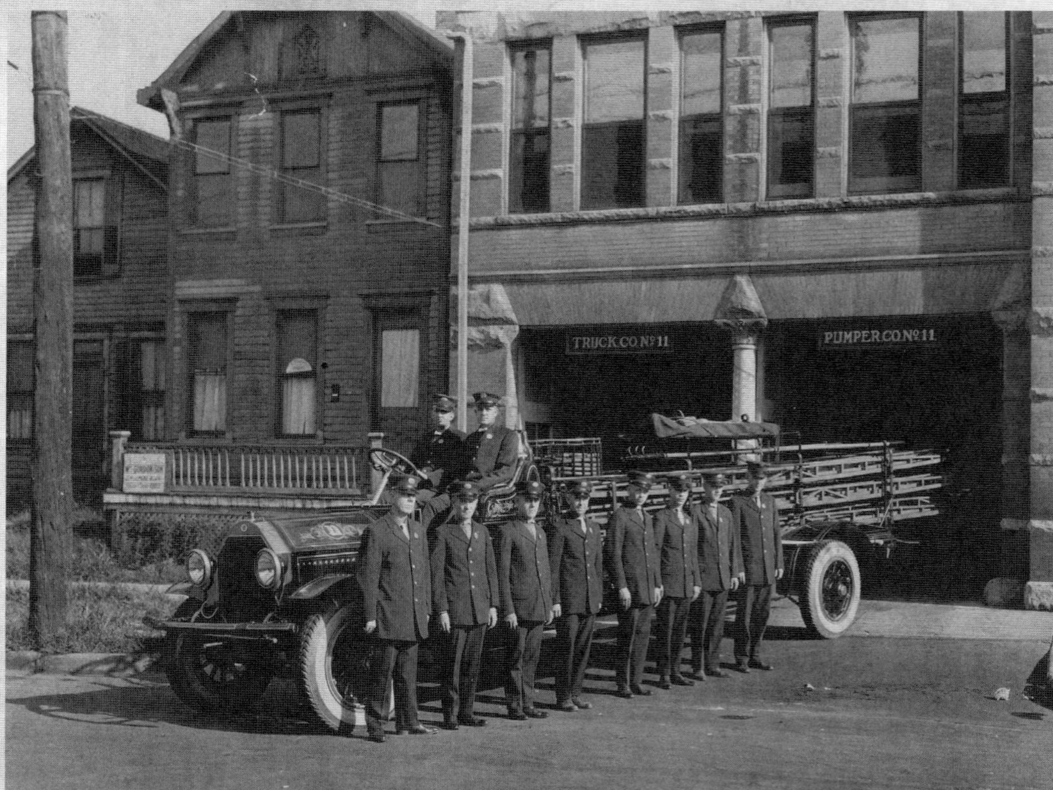

Station #11 at 1032 E. Washington Street.

Seeing the need for more safety regulations the Chief made fire prevention its own bureau and thus abolishing the Cellar Inspector in 1918. This came a year after Indiana Firemen's Association was established (1919).

Seven years after the first change in the departments working hours it happened again in 1919. The IFD wanted to add a second shift but got every fifth day off instead of every seventh day. If the firemen weren't happy with the shortened schedule they wouldn't have to wait long for it to change. Two years later in 1921 the IFD went to two shifts, a 10-hour day shift and 14-hour night shift, which would rotate every two weeks. It changed again the next year to 24 hours on 24 hours off. It would stay this way until 1928 when the Chief rotated an extra day off, lowering the work week from 72 to 68½ hours.

(See following text on page 48)

1912
Firefighters got 1 day off a week.

1912
Ladder #11 went into service at Station #7 and Ladder #17 went into service.

1912
Station #28 was the 1st Station built for motorized apparatus.

Station #30.

The Board of Public Safety in 1927 adopted new uniforms and regulations to keep the department looking uniform. Department members were to wear their class "A" uniform to and from work. This would also allow them to ride public buses and the trolley/train to work for free. Once at work, firemen would change into their "salt and peppers", which was a tweed-like material that was thick and looked like salt and pepper. These were to be worn while performing all house work, maintenance and other work at the firehouse. Then at 10 o'clock they would put on their class "B" uniform, which was a dark blue shirt and dark pants until 6 o'clock when they would change

Multiple Alarm Troy's Restaurant Fire.

back to their "salt and peppers". The department would also adopt service strips to distinguish some members for time on the job as well as identifying engineers. As the department was updating its look, they were also getting cutting edge equipment. The IFD would purchase its first breathing apparatus in 1928 called the "All Service Gas Masks", which were put on Squad 7 and Squad 30. A breathing apparatus wasn't the only new thing for the department. Since going motorized, the department would need a different kind of repair shop for the apparatus. This led to the opening of a new repair department at 1024 Saunders Street which used to be an old city barn in 1929.

1913
Squad Wagon #2 went into service.

1913
Water Tower #1 was moved to Kentucky Avenue and Maryland Street.

1914
Chief Charles Coots had the longest term as chief at 13 years.

1914
Engine #13 and #17 went into service.

1915
Engine #10, #30, and #31 went into service.

1915
Ladder #4 went into service at Station #11, Ladder #7 went into service at Station #17, and Ladder #13 went into service at Station #22.

The advancements would continue as the department opened up a new drill school also behind Station 30. Clyde Long, a substitute firefighter, would be the first person to test the new school. The drill school would last two weeks. Over the years drill school would be changed to the title of recruit class and go from lasting two weeks to 22 weeks and cover different many different disciplines including fire tactics, hazardous materials, emergency medical treatment, and extrication to name a few. Drill school would also go from being done on your own time during the day or at night to paid Monday through Friday.

(See following text on page 55)

First Female Fighter (Appointed)

On February 3, 1925, Miss Mary Frances Flynn became the first recorded female firefighter for the Indianapolis Fire Department. Truth is, Miss Flynn was never a firefighter at all, but was the stenographer and clerk in the office of Fire Chief O'Brien. The board approved the recommendation of Chief O'Brien to appoint Miss Flynn a fireman. This appointment to title of fireman also came with a raise in her yearly salary from $1,300 to $1551 a year matching that of a first year firefighter.

1916

Engine #11 went into service.

1916

Hook and Ladder Truck Company #2 reopened.

1916

Hook and Ladder Truck Company #8 and #10 went into service.

PLAZA HOTEL
LAZA HOTEL

CASH & CARRY PAPER CO.
SCHOOL SUPPLIES
CASH & CARRY PAPER CO.
SMULY
RESCUE SQUAD
No. 7 - I.F.D.

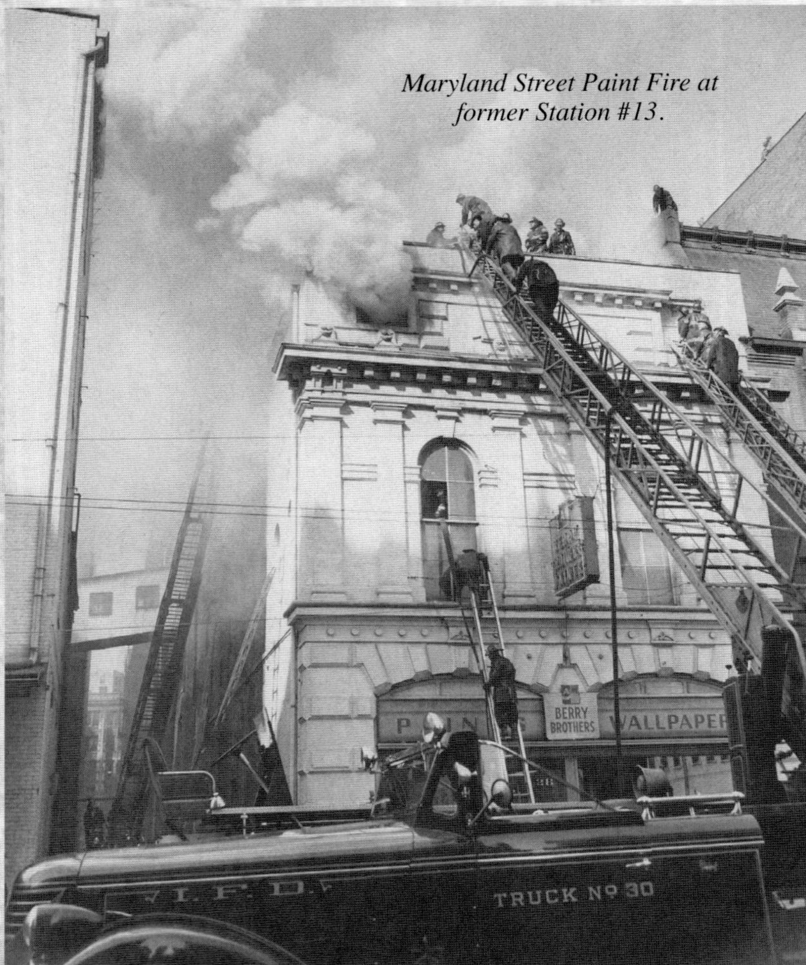

Maryland Street Paint Fire at former Station #13.

1917
Engine #1 and #22 went into service.

1917
Hook and Ladder Truck Company #12 went into service.

1917
Hose Reel #29 went into service.

1917
The Indiana Firemen's Association was established.

1918
Hook and Ladder Truck Company #13 went into service.

1918
Hose Reel #30 and #31 went into service.

Indiana State Fairgrounds

1918

The Bureau of Fire Prevention was established.

1918

The position of Cellar Inspector was abolished.

1919

The second work shift was voted down by the city counsel. Firefighters received every fifth day off instead of every seventh day off.

3 January 1920

The Fire Prevention Bureau was separated and became it's own division with it's own chief.

1920

Engine #3, #4, #8, #9, #12, #15, #20, #23, and #27 went into service.

Fire Prevention

Before fire prevention became its own division, the city was divided into districts and a Captain was in charge of inspections, which lead elimination of the position of Cellar Inspector. These were the men in charge of Fire Prevention:

1920-1922	Jacob H. Hinkene
1922-1924	Jacob E. Riedel
1924-1926	Virgil T. Furgason
1926-1929	Horace W. Carey
1929-1931	Virgil T. Furgason
1931-1944	Bernard Lynch
1944-1948	Michael Hyland
1948-1949	Henry A. Murray
1949-1953	Ralph Fender
1953-1957	Michael Hyland
1957-1960	Daniel P. O' Donnell
1960-1966	Charles P. Hill
1966-1968	Bernard Hanley
1968-1972	Don Bolinger
1972-1982	James T Mitchell
1982-1985	James Whitaker
1986-1992	Tim Whitaker
1992-1993	Curtis Gregory
1993-1997	Charles Williams
1997-2003	Dave Harris
2003-2009	Earnest Malone
2009-Present	Tandy Brauhnam

Despite all the advancements and hard work, the *Indianapolis Times* would bombard the department about their disciplinary practices and efficiencies within the department. When looking at all the advancements in cutting down response times with motorized apparatus, breathing apparatus, better gear, and better training and training facilities, this seemed preposterous as far as efficiencies go.

N. Meridian Street

In the twenties the department made much advancement like in schedules, apparatus, uniforms, rules, etc., but by the end of the decade the IFD, like the rest of the country, would be affected by the Great Depression. The IFD would show its true colors by donating 1% of all firemen's pay to assist in poor relief through the depression. The IFD had 100% participation in this relief effort in 1930. This was a proud moment for the firefighters.

(See following text on page 60)

1920

Starting in 1920, Ladder companies were moved to stations that matched their given numbers.

1920

The estimated Indianapolis population was 314,194 people.

10 December 1921

Engine #7 and Engine #13 collided at New York Street and Senate Avenue seriously injuring four firefighters.

DENTISTS

MARION PAINT CO
4 STOR
SHEET METAL
WORK

SCBA

Breathing smoke in hazardous conditions has been an obstacle for firefighters since firefighting began. In the early 1900's, firefighters were referred to as "wet beards" which was derived from firefighters wetting down a bandana and tying it around their nose and mouth to resemble bandits from the "Wild West" when entering a burning structure. It's also believed that pioneer firefighters grew their beards out so they could fold them over their mouths and nose while other firefighters grew handle bar mustaches which they shoved up their noses to keep the smoke out.

As time went on, a professional appearance was required by fire departments throughout the United States, and IFD was no different. Facial hair was limited making it more likely for firefighters to take in more smoke. The firefighters had to get creative on how to get their next breath of fresh air. These smoke-eaters would knock out windows to ventilate the room, keep their noses close to the floor, put their noses right next to the nozzle that was throwing water, and even head straight for the refrigerator door to get their next breath of cleaner air.

Then on August 11, 1928, IFD bought its first SCBA in the form of the All-Service gas masks, which were placed on Squad 7 first followed by Squad 30. Over the next 45 years many different types and quantities of breathing devices were purchased by the department. In 1931, 10 Burell All Service gas masks were purchased from Mine Safety Appliances. Six more were purchased in 1932. In 1946, two Chemox self-generating Oxygen Breathing Apparatus were purchased for $125.00. Many different types were purchased and tested by firefighters, but many presented functionality issues for firefighters ranging from being bulky in size to having an oxygen cable going back out of the structure to a pump.

In 1973, the department made a big push towards putting firefighters on air when enough MSA SCBA's were purchased so that that least one was on every apparatus on the department. Not only was this a slow and expensive process for the department, but it was one that required a change in mind-set for firefighters. The smoke-eating days were coming to an end. In 1974, the department got rid of the Chemox masks because the firefighters chose to use them over the safer MSA's because they weighed less. Over the next several years, enough SCBA's were purchased so that each position on the apparatuses had them. To service the masks, in 1976 the department opened up the "White Room" at the repair shops. In 1984, the department went to Scott's 2.2 SCBA. A year later, each firefighter was issued their own face piece. This allowed firefighters to maintain better hygiene and eliminated the passing of germs and bacteria back and forth among firefighters. In 1994, the department switched to Interspiro's. The department has made it a habit to review the SCBA technology every ten years or so, making sure the firefighters have the safest and most affordable product of SCBA. In 2002 the department changed to their current SCBA, Scott's 4.5.

1921

Engine #2, #6, #14, #16, #18, #19, #21, #24, #25, #26, and #29 went into service.

1921

Ladder #1, #2, #3, #5, #14, #18, #19, #22, #25, #27, #29, and #30 went into service.

1921

Ladder #4 closed.

1921

Ladder #7 was moved to Station #7, Ladder #11 was moved to Station #11 and Ladder #13 was moved to Station #13.

1921

The second work shift was introduced. Firefighters worked the 1st 10 hours or the second 14 hours for two weeks and than would swap schedules.

June 1921

The department became completely motorized as the last horse was retired.

3 September 1922

Engine #31 closed.

4 October 1922

Engine #32 went into service.

1922

African-American Hose Company #9 was moved to 441 Indiana Avenue and renumbered Hose Company #1 at Station #1. It was the only station where African-Americans could serve. African-American firefighters were only hired to replaced African-American firefighters who retired or died at Station #1 up until 1960.

1922

The work schedule went to shifts of 24 hours on and 24 hours off.

1 June 1925

Ladder #32 went into service.

1927

Squad #7 and #30 went into service.

5 April 1928

The Board of Public Safety adopted a regulation for uniforms specifying the uniform for station wear, helmets, rubber coats, and boots.

28 July 1928

Engine #31 reopened.

1928

An all service gas mask was put into service on Squad #7 and Squad #30.

1928

The work week was lowered from 72 hours to 68 ½ hours rotating an extra day off for firefighters.

Feeling the pressure to save money and do more in the 1930's, the department started to build their own apparatus called "shop builts" in 1935 saving the city money. They also hired civilians to work as clerks for the chief instead of merit firemen for the first time in 1935. Some new technologies were introduced such as a mobile radio put in the chief's car in 1932. A red light system was installed in the Gamewell system in 1934 turning all downtown lights flashing red while the department was responding on a run.

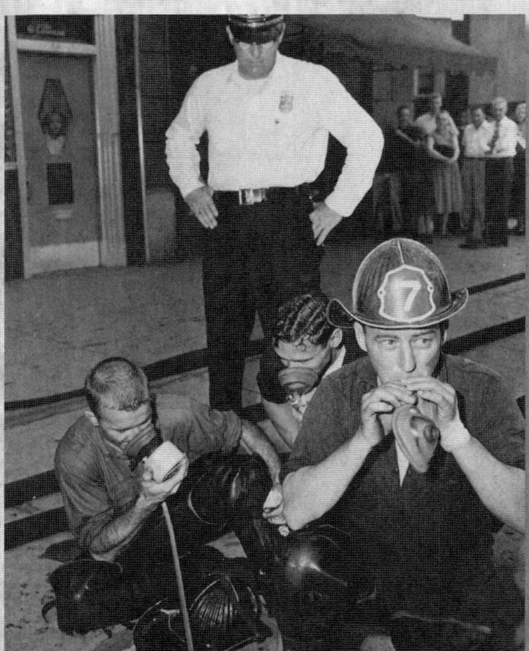

Shop Builts

In the mid 1930's when the country was experiencing the worst of the Great Depression, IFD took it upon themselves to reduce operation costs by building their own fire apparatus. The shop builts, referred to as "IFD Specials" on the streets, turned out being some of the most reliable pieces of equipment with some staying in service on the streets well into the 1950's.

The IFD would start doing more in the form of EMS although they technically didn't have a division or do first response at this time. Squad 7 would receive a K&J Resuscitator, which could be used for cardiac arrests and for fire victims. This was a pretty big purchase financially and technologically for the IFD that would some years later lead to doing EMS.

Fire at Illinois Street and Vermont Street.

6 February 1929

Fire Prevention inspectors got regulation hats, coats, and badges.

15 February 1929

A new mimeograph machine were placed in the chief's office.

1 November 1929

The new repair department opened at an old city barn located at 1024 Sanders Street, located behind Station #30.

1929

Clyde Long was the first substitute firefighter sent to a drill school for two weeks.

1929

Service stripes were adopted for members.

1929

The new drill school opened at Station #30.

Although the department had built a training facility a few years prior, it didn't host its first drill school until 1935 when the IFD would start numbering its recruit classes. A year later the department would have its first merit promotion process for lieutenants and captains. The first merit promotion process for chiefs would come in 1939. There was also a school to go to that would help applicants learn what the roles and responsibilities were of a fireman as well as educate them on department guidelines, rules, and tactics. This was a change brought about by a new merit and charter system. This changed the department forever from the "good ol' boys system" to helping level the playing field for applicants and promotions.

(See following text on page 66)

20 November 1930
The 1st 1 1/2" hose was introduced to department.

1930
The department unanimously voted on a 1% assessment of wages for poor relief.

1930
The estimated Indianapolis population was 363,161 people.

1930
The Indianapolis Times starting bombarding the department in regard to discipline and efficiency within the department.

Recruit Classes

In 1935, the department started assigning recruit classes a number in order to track the future firefighters of the department. At the time, firefighters would go to two weeks of training classes on their own time, unpaid, and then would be placed on a list to be hired at a later date. The recruit classes below are when the firefighters attended those training classes. This practice would continue until 1969 when firefighters got paid for training, making the date for the training classes and their hire date the same date.

1 – July 29, 1935
2 – July 13, 1936
3 – November 10, 1936
4 – December 11, 1936
5 – April 19, 1937
6 – November 17, 1937
7 – April 3, 1939
8 – November 12, 1940
9 – July 10, 1942
10 – March 25, 1946
11 – October 10, 1946
12 – November 3, 1947
13 – June 22, 1948
14 – 1949
15 – July 5, 1950
16 – January 16, 1951
17 – August 10, 1951
18 – October 17, 1951
19 – May 5, 1952
20 – November 9, 1953
21 – July 6, 1954
22 – January 1955
23 – September 1955
24 – January 1956
25 – September 1, 1956
26 – 1957
27 – January 1959
28 – May 1960
29 – Mid 1961
30 – Early 1962
31 – September 1962
32 – July 1966
33 – October 1966
34 – 1967
35 – April 10, 1968
36 – December 18, 1968
37 – July 9, 1969 *
38 – October 1, 1969
39 – January 7, 1970
40 – October 14, 1970
41 – July 7, 1971
42 – November 24, 1971
43 – October 11, 1972
44 – October 10, 1973
45 – January 16, 1974
46 – July 2, 1975
47 – November 5, 1975
48 – June 16, 1976
49 – January 12, 1977
50 – March 3, 1978**
51 – May 11, 1978
52 – January 19, 1979
53 – November 19, 1979
54 – November 3, 1980
55 – April 19, 1983
56 – March 12, 1984
57 – March 7, 1985
58 – February 25, 1986
59 – August 12, 1986
60 – February 25, 1987
61 – June 2, 1988
62 – November 27, 1989
63 – July 6, 1990
64 – March 4, 1991
65 – January 3, 1995
66 – October 5, 1995
67 – May 6, 1996
68 – December 16, 1996
69 – February 23, 1998
70 – March 8, 1999
71 – March 20, 2000
72 – April 9, 2001
73 – January 21, 2003
74 – February 14, 2005
75 – February 13, 2006
76 – March 19, 2007
77 – June 9, 2008

*1st paid class
**1st Merit female, 1st EMT

Foxworthy Ford Fire

Central Station Fire

PIZZA
Nº
30
Nº
30

14 December 1931

The Indianapolis Fire Department received 10 Burrell All service gas masks.

1931

All engine companies were equipped with 200' of 1½" hose lines.

1931

One K and J Resuscitator - Inhalator was placed into service on Squad #7 costing $375.00.

27 February 1932

The Indianapolis Fire Department received six additional Burrell All service gas masks.

11 March 1932

The cities remaining 40 cisterns were dismantled.

14 March 1932

Squad #14 went into service for a 60-day trial period.

Towards the end of the 1930's the department, as well as the country, was starting to get back on its feet and on track, but it wouldn't be for long. The start of World War II would affect the IFD like most of the country. By the end of World War II, 97 of the IFD's bravest fought for our country, losing one in the process. With so many of the departments men fighting over seas the department on October 16, 1943, would start working "duration shift", meaning they were working for men at war for free paying no overtime. While working duration shifts the firemen were going through uniforms more frequently, and with that, started to receive a clothing allowance of $60.00 in two payments of $30.00 with the first being received on August 17, 1943. The department was running short and tired, for their tireless efforts for each other and the city they received some benefits. One in the form of the chief's aide being promoted to lieutenant (1943) and the firemen received a Kelly day which meant every two weeks they would get three days off (1948).

The Indianapolis Times Clothe-A-Child, Mile-O-Dimes, & Fill the Bell

In the fall of 1930, while reporting about two small shanty communities along the banks of the White River, a reporter made mention of the children in these two communities that were completely underdressed for the upcoming winter. On December 1, 1930, an article was printed in the *Indianapolis Times* suggesting that people who wanted to donate could do so by clothing these children by buying hats, gloves, shoes, pants, shirts, jackets, etc. In this first year clothing was provided for 422 children.

On December 13, 1933, the first Miles of Dimes (later Mile-O-Dimes) opened along Washington Street in front of L. S. Ayres and S. S. Kresge Building. People would donate dimes, and other coins, by placing them in rows on the side walk. Policemen and plain-clothed men would stand over the coins 24 hours a day until Christmas Eve when the coins were swept up. The first year it measured 1,912 feet and raised $3, 207. 55. The money raised was donated to the Clothe-A-Child program.

In 1946, members of the Firemen's Post 42 of the American Legion took over the role of standing guard over the money on the sidewalk. Members, as well as off-duty firefighters, would watch over the money until Christmas Eve, which continued until 1958. In 1957, many charities drives, including the Clothe-A-Child campaign, were being brought under the auspices of the United Fund to prevent duplication. The sidewalk Mile-O-Dimes was replaced by a booth. In 1960, The *Indianapolis Times* gave up running the Clothe-A-Child program and within a few years closed up shop as well.

In the 1970s, the Indianapolis Fire Department restarted the "Fill the Bell" and "Cloth-A-Child" programs that still exist today. Firefighters stand the bell in local department stores in two hour shifts dressed in class "A" uniforms collecting money for the annual Cloth-A-Child event.

The "Helmet"

As a way to communicate to the membership, the Local 416 Union started a publication, *The Helmet*, in 1946. This monthly booklet was filled with various information including department information, the announcement of births for firefighter's families, results from local bowling tournaments, and even sometimes great recipes for the fire house. In February 1963, the publication was suspended until 1966 when it was started back up. Today the content ranges from run totals for individual apparatus to current department information. *The Helmet* can be found floating around the fire houses, in hard copy, or can be accessed through the l416 website by the membership.

28 July 1932

The 1st radio was installed in Chief Voshell's car.

19 December 1932

Squad #30 closed.

1933

Foam Car #13 and Hose Tender #13 went into service.

The department continued looking at new advancements in breathing apparatus this time purchasing two Chemox Self Generating Oxygen Breathing Apparatus. Two-way radios with the call letters WHRU went into service in 1948. It would only stay like that for two years then they would change to KSA890. Before the radio advancement, the department started a publication called *The Helmet* in 1946 as a way for the Chief and members to keep up with department and personal news. This now monthly publication told of department events, news, marriages, child births, family deaths, the IFD sporting news, jokes and stories.

(See following text on page 74)

2 June 1934
District #1 was moved from Station #1 to Station #5.

26 June 1934
A red light system was installed at the Gamewell system turning all lights west of downtown "flashing red" giving emergency vehicles the right of way.

12 February 1935
Engine #31 re-closed.

11 June 1935
The Indianapolis Fire Department built their apparatus from the ground-up called "Shop Builts". Truck #5 and Engine #7 were 1st of 9 apparatus to be built.

29 July 1935
The 1st school for drill school was held.

1935
Engine #2 ran out of Reel House #9's quarters until 1937.

1935
Engine #6 closed.

Local Union 416

On October 4, 1934, the IAFF Indianapolis Local 416 Union was started. Charter officers Elwood Johnson, John E. Doyle, Michael Hiland, Calvin Martin, Harry C. Atkins, and Harry A. Miller got together to come up with a fair way to represent Indianapolis firefighters during a period when firefighters were grossly underpaid during the depression years.

Through the years, the role of the Union 416 grew. During the 1960's when firefighters had accidents, city legal didn't always show up to represent the engineers. Union 416 hired attorney Jack Ruckelshaus to represent IFD and the Fraternal Order of Police (FOP).

In 1966 the Union helped negotiate with the city to lower the work week to 56 hours and bring on a third shift of firefighters. In 1968, the Union made headlines when there was pay raise dispute with Mayor Richard Lugar. Union President Maynard Byrum told the *Indianapolis Star* that 284 out of 878 firefighters were eligible to retire and would do so if an agreement wasn't reached between the Union and the city. The raise was granted. Ten years later the Union wanted their members to be view as professionals in the work place, which resulted in the name change from "Indianapolis Firefighters Association" to "Indianapolis Professional Firefighters Union" in 1976.

In addition to negotiating contracts with the city and representing firefighters on disciplinary issues, the Union gained parity with FOP. In 1977, the FOP negotiated a $2000 pay increase to all police officers because some of the officers were losing their take home police cars. The Union and Mayor Bill Hudnut couldn't come to an agreement and against the advice of other IAFF Unions, the Union 416 decided to go on strike with over 90% voting yes to strike. On the day that the strike was to occur, with 25 minutes to spare, Mayor Hudnut agreed to federal mediation averting the strike. As a result, the Union created the Political Action Committee (PAC) and became involved in politics from township trustees to governor.

In the 1960's the Printer's Union allowed the membership to meet in their hall located at 315 North Delaware Street. The executive work was done at home or at the fire house until 1970 when the first office was rented above the Hawkins Pharmacy at 245 East Southern Avenue. As the Union grew, so did the need for an office and meeting hall. In the early 70's, the building at 3701 West 10th Street, which housed the Western Apparel Shop, was purchased for $25,000. Over the next decade, many parties and meetings were held in this small hall located just down the street from the Indianapolis Motor Speedway. The hall would open up the week prior to the Indianapolis 500 where firefighters from around the world would pop in for a drink or two. The Union remained in this building until the early 90's when it moved into the current location. But the current location was not an easy transition for the Union.

The current building at 748 North Massachusetts Avenue was discovered in 1984 by the Union as a possible place to relocate. The building had already played a role in IFD's history as Station #2 from January 1, 1872 to October 18, 1895 and then renumbered to Station #8 until February 2, 1932 when it was closed. This deplorable building had seen many different tenants as well as many different fires including a second alarm fire in the 70's. The condition of the building raised some concern from the members and it was agreed upon that no union dues would be used to purchase, renovate, or remodel the site. The money collection process began. The Union received grant monies from several places including the Lilly Foundation and the City of Indianapolis. The members held many fundraisers, Back on Track parties in May, Halloween parties in October, hockey tournaments, and reverse raffles, just to name few. With the increase of cash coming in, the remodeling and acquisition continued as well. There was purchase of the building next door, the old livery stable, which was later connected by a skylight which doubled the size of the hall. Then there was purchases of the three houses as well as the lots to the West and the Massachusetts Avenue point to the east. All the monies collected went directly into these purchases and upgrades. During a visit to the Mishawaka Local, President Mike Schenk, found a unique program called Survive Alive. This educational program was created to involve the largest number of young children in the fire district as possible to learn about fire prevention and life safety strategies. With this vision in place, the funds started to roll in from many corporations and individuals.

In March of 1993, the Union offices opened in the front of the building. At the time the building still lacked plumbing. The old hall was sold bringing additional monies for the upgrades. In 1994 Survive Alive opened educating second graders throughout the city. Over the years the upgrades continued with the opening of the Warren J. C. Smith classroom and the Survive Alive village.

After the tragic fire at the Athletic club that took the lives of John Lorenzano and Woody Gelenius, a memorial service was held for all fallen firefighters in Indianapolis. One of the daughters of the fallen firefighters approached the Union Treasurer and thanked him for recognizing her father's sacrifice. This started the call for the fitting memorial and tribute to all of our fallen members. Once the memorial was designed, personalized bricks were sold around the memorial raising nearly $50,000. In July 1996, the memorial was dedicated in grand ceremony and this began the annual Reading of the Names tradition that takes place with the start of Fire Prevention Week in October.

The Museum opened the same year. Members spent countless hours painting and preparing the museum. Artifacts and donations of fire department memorabilia started coming in including three horse drawn apparatus that were received from the Indianapolis Motor Speedway Museum through the generosity of the Hulman-George family. A plaque from headquarters was moved to the museum memorializing the sacrifices made by firefighters in WWII. With many firefighters serving in wars for the US, the Veteran's Walk of Distinction was established and dedicated in May of 2000.

Over the years, the Union has established many traditions including the Children's Christmas Party and Easter Egg Hunt, retiree's nights, the pumper pull and party during FDIC week, and the St. Patrick's Day party that follows the parade. They host many fundraisers and privates parties including Media Stew Nights as well as sponsoring racing teams for the long running of the annual Indianapolis Mini-Marathon race. The Local 416 Union is dedicated to representing the firefighters of Indianapolis.

1935

The 1st civilian office employees were hired as merit clerks in the chief's office.

1935

The 1st department Trial Board was established to discipline firefighters having unpaid bills.

1936

The 1st merit promotion school for Lieutenants to become Captains was held.

26 January 1937

A total of 52 firefighters were sent to Evansville, Indiana to assist with flood relief. An additional 27 firefighters were sent 3 days later. While assisting with flood relief, the department suffered its 1st Line of Duty Death (LODD) outside of Indianapolis.

23 May 1938

Squad #14 went into service.

6 June 1938

Engine #17 ran out of Station #30 and Ladder #17 out of Station #4 until October 17, 1941.

6 June 1938

Station #17 closed.

1 August 1939

Ladder #30 closed and the apparatus was used for Ladder #13.

1939

The 1st merit promotion school for Captains to become Chiefs was held.

1939

The Indianapolis Fire Department had a travel baseball team. They eventually played in Wrigley Field in Chicago.

Ladder Company #13.

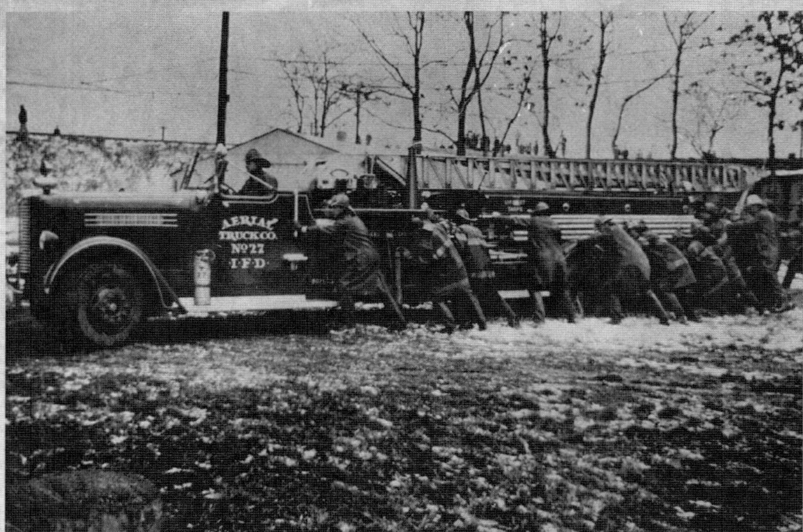

Cinema Art Theatre Fire

1940

The estimated Indianapolis population was 386,972 people.

16 January 1941

The 1st firefighter was activated by the selective services.

4 September 1941

Ladder #30 reopened.

3 November 1941

The 1st Chauffeurs school was held.

16 October 1941

Engine #10 ran out of Station #17 until October 15, 1952.

17 October 1941

Station #17 reopened.

1941

A total of 97 firefighters served our country in World War II.

Murat Firemen's Club

On January 6, 1949, The Murat Firemen's Club held its first meeting in the Kneipe Room of the Murat Temple electing firefighter Ray Cogan as the first President. Beginning as a marching club representing the Murat Shrine and firefighters, in 1980 the club became motorized. The objective of the club is to provide Murat Shrine with a Safety Committee and form a social club for Indianapolis Indiana firefighters, active and retired, that have a common interests and are members of Murat Shrine.

As they entered the 1950's at time of peace, family values, rock n' roll, and fun shaped the country. The department would expand it's minority staffing and community awareness adding its first African American truck company on March 1, 1951. That wouldn't be the only expansion though as the IFD would add two battalion chiefs, District 5 on January 1, 1951 and District 6 on March 15, 1955. In addition, in 1955 the IFD would install an ultra-modern communication system in headquarters and would now be referred as the "Alarm Office." All the updates, training, and hard work was not going unnoticed by the NFPA. In 1953 Indianapolis had the 3rd ranked class "2" fire department in the country as well as won

honorable mention as one of the best departments in the country for all same size cities. By this point the IFD had won 16 awards in a five-year span. By 1956 100% of the IFD would be trained in first-aid and members would start to receive longevity pay, which is an extra stipend of money per year of service. In this decade the department would team up with Muscular Dystrophy to help raise money, this continued the departments' high priority for charity work and outreach.

17 October 1942

Engine #20 ran out of Station #12 until the new station opened on September 18, 1952.

1 January 1943

Battalion Chief aides received the rank of Lieutenant.

17 August 1943

A clothing allowance of $30.00 was given to firefighters.

16 October 1943

Firefighters begin working duration shifts to fill vacancies on both shifts.

Training Directors

May 1922	Herbert H. Fulmer
January 1935	Harry Voshell
January 1940	Audray Craig
October 1949	Herbert Dwyer
August 1962	Kenneth Williams
January 1967	Sam A. Oslos
January 1972	Richard G. Harding
July 1981-April 1982	Richard VanSant Sr.
April 1982-June 1984	Ronald Callahan
July 1984-February 1986	Harry Tibbetts
March 1988-July 1989	Warren White
November 1989-November 1991	Greg Roembke
March 1992-1995	Chris Pitts
1995-1997	Joe Krebsbach
1997-2000	Rick Longerich
September 2000-March 2002	Steve Dixon
June 2002-July 2004	Doug Abernathy
July 2004-January 2006	Paul Sullivan
January 2006-2007	Phil Reuter
October 2007-April 2008	Vince Vaughn
June 2008-Present	Charlie Fadale

Canary Cottage Fire

8 April 1945

Charles C. Haas became the 1st and only firefighter who lost his life in World War II.

25 May 1946

The Indianapolis Fire Department received 2 Chemox Self Generating Oxygen Breathing Apparatus.

1946

The 1st Issue of the "Helmet" was published by Arnold Phillips.

1946

Three active firefighters and one retired firefighter pled guilty to a ticket scam at the Indianapolis Motor Speedway.

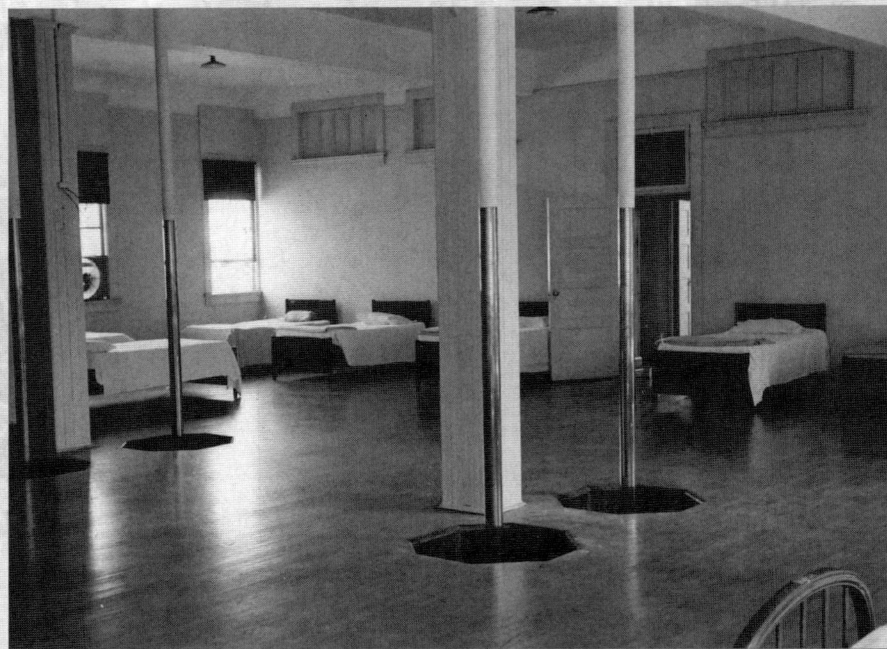

Station #7 Bunk Room

S. Capitol Avenue and McCarty Street

1 January 1948

Two-way radios with call letters WHRU went into service.

1948

All stations had to hit the "OK" button in house to confirm getting a run.

1948

Firefighters were given three days off every two weeks while working a 24 hours on - 24 hours off schedule which was called the "Kelly Day".

1948

State Bill #283 was passed requiring Chiefs to have a minimum of 5 years of active service, and 2 years experience for all other promotions.

6 January 1949
The Murat Fireman's Club was started.

26 April 1949
Ladder #3 closed and reopened as Ladder #15.

1949
Department staffing consisted of 711 firefighters.

1949
Shortest run to cover a plug was 7 feet by Engine #17 when Box 58 at their firehouse was pulled.

1949
The city house passed a uniform bill giving free uniforms to firefighters and police and an additional $100.00 for supplies.

Firefighter's Credit Union

On March 1, 1956, a meeting was held at old fire headquarters at 301 East New York Street to bring together a cooperative venture to help meet the financial needs of firefighters and their families. Through the years, many other volunteers have given an enormous amount of time and effort to the credit union. In 1978, the building at 501 North New Jersey Street was purchased. Since the purchase, many modifications, upgrades, and an expansion has made the credit union what it is today. With a current staff of 23 employees, the birth of the Indianapolis Firefighter's Credit Union has created many opportunities for firefighters and their families financially including general banking needs, home and car loans, as well as education loans.

1950

In the 1950's, IFD had 157 2nd alarm or better fires, an average of 15.7 a year.

1950

The drill school was moved to 1445 West Michigan Street and was renamed the Fire Academy.

1950

The estimated Indianapolis population was 427,173 people.

1950

Two-Way radios changed its call letters to KSA890.

1 January 1951

District #5 went into service at Station #31.

1 March 1951

The department had its 1st African-American Truck Company.

1951

Department staffing consisted of 749 firefighters.

3 April 1952

The 1st school for drill instructors was held.

10 April 1952

Ladder #11 closed and reopened as Ladder #21.

15 October1952

Engine #10 closed and reopened as Engine #31.

1952

Engine #6 reopened.

16 January 1953

Pay days were changed from the 5th and the 20th of each month every other Friday.

21 April 1954
Ladder #30 re-closed and reopened as Ladder #31.

21 April 1954
Engine #30 ran out of Station #6 until October 9, 1958.

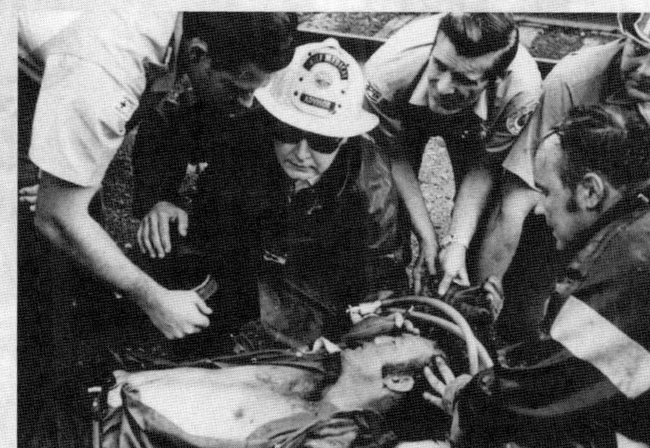

15 March 1955
District #6 went into service at Station #20.

3 December 1955
The new ultra modern communication equipment was put into service "Alarm Office" at Headquarters.

1955
Station meals on average cost from $.55 - $.75 for both lunch and dinner.

1956
100% of the department completed Red Cross 1st aid training.

1956
The department started paying longevity pay for years of service on the department.

22 December 1957
The 1st year that the Indianapolis Fire Department and Muscular Dystrophy teamed up.

1957
The Hundred Club was established.

25 August 1958
Engine #10 Reopens.

25 August 1958
Ladder #21 closed and reopened as Ladder #10.

Interstate Warehouse Fire on S. Missouri Street.

9 October 1958
Ladder #30 reopened.

1958
The Marion County Rural Fire Dispatch was started.

1959
The department upgrades Squad's #7, #14, and #17 to carry 200 gallons of water.

Station #25

IFD Sports

Over the years, IFD has been well represented by firefighters participating in many different sports. Whether it was the bowling leagues in the early 1920's & 1930's that yielded monies and local recognition in the newspaper, IFD's baseball team going head-to-head against Chicago Fire Department's baseball team in Wrigley Field on September 12, 1938, or the softball players of the annual Kriech Softball tournament, firefighters have found ways to stay active with their co-workers outside of the fire house. Additional sports have included Baseball, Basketball, Bop-To-The-Top teams – a stair climbing competition held downtown Indianapolis, Flag Football, Football, Go-Karts, Golf, Hockey, Mini-marathon running, Rugby, and Scuba diving.

1300 Block of W. 29th Street

CINEMA
CINEMA
ART THEATRE
RADIO
Theatre
ANTIQUES
5

A Paradigm Shift 1960 - 1985

The IFD reflected the issues of the country in the sixties as well. The sixties were crowned by the civil rights and free love. The department seemingly one step ahead, desegregated in 1960 with full integration. It is unclear how welcomed this was, but was a much needed evolution. In this time the department would promote its first African American chief; Chief Pelman Johnson. The IFD held with the utmost respect in the community and a following of citizens who wanted to serve found the need and want to form the Indianapolis Fire Buffs. This is an organization of citizens that assist the IFD with rehab work, charity, fund raising, and anything asked or felt they need to do to help the department. They formed in 1962. The following year *The Helmet* stopped running. This would be a short-lived occurrence as it would once again start in 1966

End of Segregation

On January 1, 1960, segregation in the Indianapolis Fire Department officially ended. Prior to this date, African-American firefighters were only hired to replace retiring African-American firefighters and were only assigned to Station #1. With the "silent treatment" and a series of practical jokes towards some African-American firefighters from some Caucasian firefighters, it would be some time before the race barrier was breached.

1 January 1960

Segregation officially ended within the Indianapolis Fire Department with the implementation of full integration.

1960

A chief's salary was $9,300.00 and a private's salary was $4,600.00.

1960

Engine #22 was detailed to the State Fair and answered nine fire alarms.

1960

In the 1960's, IFD had 175 2nd alarm or better fires, an average of 17.5 a year.

1960

The estimated Indianapolis population was 476,258 people.

Coliseum Explosion

On Halloween night in 1963, an explosion rocked the State Fairgrounds Coliseum killing 74 people and injuring nearly 400. A packed coliseum of 4000 people were attending an ice show when a leaking propane tank exploded, sending people and debris flying across the arena. At 11:06 p.m. the dispatcher struck Box 2137 to the fairgrounds for an explosion. A second explosion moments later created a large fire ball devastating everything in its path. Firefighters, the police, the Red Cross, the Salvation Army, and many volunteers helped with the rescue and recovery of victims eventually turning the ice rink floor into a make-shift morgue until the bodies could be identified.

and would continue to run up to the present. 1963 would be scared in the city's mind as an explosion at the Coliseum rocked Indianapolis killing 75 and injuring 400. The department would use the ice rink as a make shift triage station while sifting through the rubble for victims.

This time period also continued the IFD's strong commitment to charity as it started Fill the Bell in 1966 and raised $2542.00. Fill the Bell raised money to clothe underdressed child in the city. The department would also expand its command staff adding a Personal and Public Relations Department. The IFD would need these positions, as fires would rise significantly due to civil unrest and the city buying properties to install two new interstates through downtown. The city would put the interstate closer to its center than any other city in the nation. Depreciated property value and urban plight would mix to cause a rash of fires that would put some of the IFD's apparatus among the busiest firefighting apparatus in the country. Many downtown fire companies were absolutely exhausted at the end of their shifts going from fire to fire. Some firemen reflect on some of their shifts of "pickin-up" (loading hose and equipment up after a fire) at a fire and seeing a plumb of smoke in the air just a couple of blocks away and heading to that fire. This rapid pace of

(See following text on page 93)

Coliseum Triage

Indianapolis Fire Buffs

On May 18, 1962 Chief Arnold Phillips assembled a meeting at his house with five men he was acquainted with that had the welfare of the IFD personnel in mind. He had recognized these individuals on the scene of many fires with thermos full of coffee for tired firefighters. This was the start of the buffs with their canteen. Currently the buffs hold meetings, raffles, and host a breakfast at the fire house for the firefighters. On many working incidents you will find the buff's providing drinks and snacks for firefighters in one of their Rehab Support Units (RSU).

1962
The Indianapolis Fire Buffs was organized.

1 February 1963
The "Helmet" was suspended until 1966.

10 July 1963
The State Fair Coliseum explosion killed 75 people and injured an additional 400 people.

1963
Station meals on average cost from $1.00 to $1.50 for both lunch and dinner.

26 October 1964
Engine #33 went into service.

DOVER

CABINETS
ROOFING
Lyon LUMBER AND SUPPLY CO. INC.
MILLWORK
PLASTER
Nº 27

1 May 1965

The 1st annual Firemen's Golf Tournament was held.

1965

A total of 738 fire alarm boxes were in service.

1965

A total of 9,560 fire hydrants were in service.

1965

The city was sued for $1 million in back pension expenses.

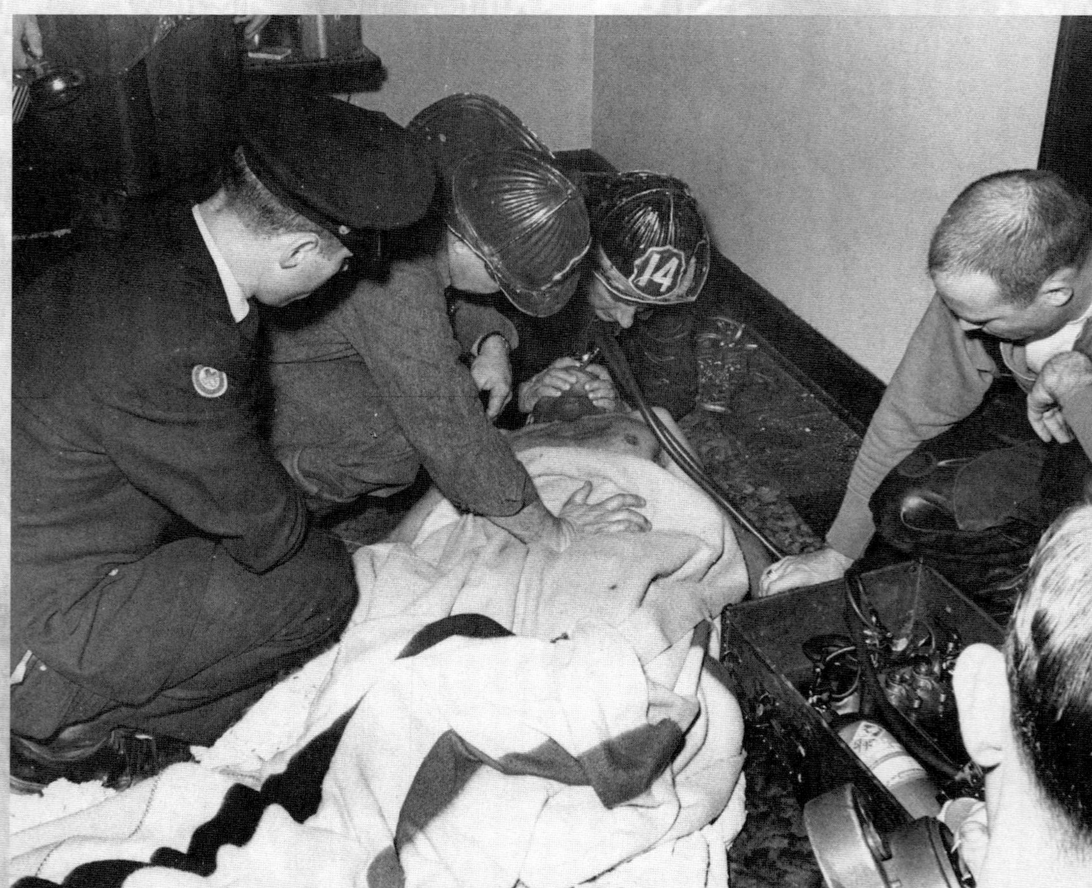

fires would help facilitate the need for a third shift. The third shift would be implemented on August 10, 1966. This would drop the hours worked from 63 to 56 per week. The schedule would go from 24 hours on 24 hours off to 24 hours on and 48 hours off. With this new shift the IFD would also change uniforms, doing away with the "salt and peppers." After some short experiments with a uniform and cleaning service the department decided on light blue class "B" shirts and dark blue pants. They also did away with wearing class "A" uniforms in to work and all the uniform changes throughout the day.

10 August 1966

The third shift was started, dropping the work week from 63 hours to 56 hours. Firefighters went from a 24-hour shift on – 24-hour shift off to 24-hour shift on – 48 hours off eliminating a lot of overtime.

1 December 1966

Fill the Bell was started collecting $2,542.00 and clothing 55 kids.

1966

Pelman Johnson was the 1st African-American District Chief.

1966

Squad #17 went into service.

26 April 1967

Engine #34 went into service.

26 April 1967

Ladder #2 closed and reopened as Ladder #34.

1967

A Life Gun launched rescue rope went in service on Squad #7.

1967

A total of 760 fire alarm boxes were in service.

1967

A total of 9,632 fire hydrants were in service.

1967

The department changed the color of city hydrants from yellow to aqua.

1 January 1968

Indiana state law mandated the retirement age at 65.

5 March 1968

The Arson Unit started with Car #10 going into service.

1 November 1968

The Personnel and Public Relations Department was created.

9 November 1968

The department received a $115,000 Jet Ranger Helicopter.

13 November 1968

The Public Information Officer (PIO) position was established with the rank of Captain.

4 December 1968

The 1st annual Police and Fireman's Recognition Day was held.

16 February 1968

The 1st 911 system was put into place in Haleyville Alabama on February 16, 1968.

Arson Directors

The Arson Division was started on March 5, 1968, with CAR 10 going into service. Over the next two years the division started inspecting buildings as well as investigating arsons. With the creation of the Arson Photo Division on January 1, 1970, a need for a director was established. In February 1980, IFD and the Indianapolis Police Department started co-staffing arson investigations. Still today, an IFD and IMPD officer work the arson cases.

William G. Earle	January 1970 – January 1977
Warren Bogard	January 1977 – March 1977
Ronald Callahan	July 1977 – May 1979
James Mitchell	May 1979 – May 1981

(Arson came under Fire Prevention, see page 117)

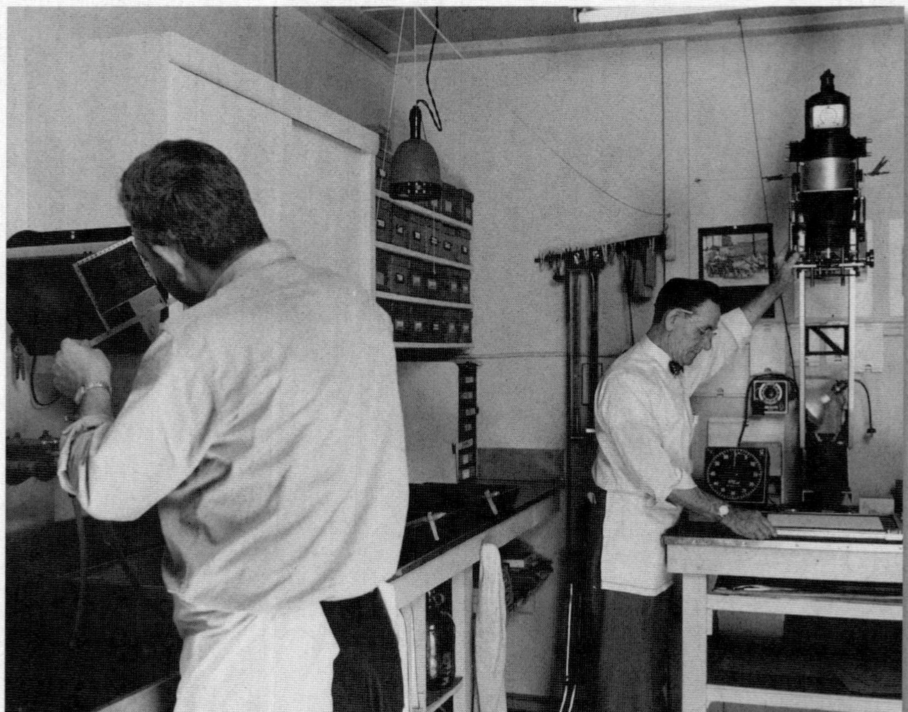

As far as business goes, the department would share a purchase with the police department of a Jet Ranger Helicopter bought in 1968 for $115,000.00. It was surprisingly used quite often. Also, the department would start to make the change from buying gasoline apparatus to diesel. The first in service was Engine 28 in 1969. It could pump longer without problems and used far less fuel as well as using a cheaper fuel, this just four years after starting to purchase cab over apparatus (1965). Equip-

Shop Built Engine #14.

ment wouldn't be the only big change at the end of the decade, as the IFD would start to have paid recruit classes. This would also change how hiring was done. Before the department would run a process and invite x number of people to start recruit school and candidates would be hired based on where they finished and how many the department needed. With this change the department would need to be x amount short to hire and then hire x amount and pay them to go to recruit school. Before the department hired one or two people at a time; now its 20-50 at a time. This would be the last big change of the decade, but one of the biggest changes was on the horizon.

1968

Department staffing consisted of 878 firefighters.

1968

Firefighter Joseph Kimbrew received the 1st "Firefighter of the Year" award. He was later appointed Chief of the Department in 1987.

1968

The department received the 1st Engine with closed cabs, which became Engine #12.

Gas Plant Fire

ROAD
CONSTRUCTION
ENDS

The seventies would start out with a bang as mayor Richard Lugar would pass one of the most important pieces of legislation in the city, as far as the fire service is concerned. This bill referred to as "UNIGOV" changed what could be done with schools, trash pick-up, law enforcement, and fire protection. Ironically though, it would not come to fruition for another 35 years. This bill changed the boundaries of the city to include the whole county doubling the population. This would catapult Indianapolis to the 11th largest city in the country with the stroke of a pen. This bill would change the way hiring, firing, promotions, demotions, and discipline where done as the merit system was changed. This would be a resonance period for Indianapolis as it was now considered a major city. The city would beef up its infrastructure doing many large projects that would take the next twenty years to fulfill; the first of which would be to add interstates starting in late 1969 to early 1970. This would put two interstates I70 & I65 right through downtown.

(See following text on page 100)

24 April 1969

Truck #7 becomes Ladder #7, with a tiller.

1969

The 1st diesel engine is used by the department, Engine #28.

1969

The 1st paid recruit class was held.

1969

The department activated Code 4 for the 1st time calling off duty firefighters to report to designated task force areas to assist due to looting at the buildings in the 700 block of Indiana Avenue which results in 82 arrests.

1 January 1970
The Arson/Photo Division was created.

1 January 1970
Uni-Gov established special fire service districts for all Marion County including Indianapolis. Also redefined the Merit system for hiring, promotions, firing, and demotions.

28 July 1970
A total of 52 firefighters were injured and 1200 residents were evacuated from a chlorine gas blaze on the city's northeast side.

1970
In the 1970's, IFD had 104 2nd alarm or better fires, an average of 10.4 a year.

1970
The estimated Indianapolis population was 744,624 people.

1970
The Explore Post 721 was created.

1970
The Indianapolis Fire Department started delivering Emergency Medical Services (EMS).

1970
The total fire loss was recorded at $9,076,409.00.

Midstate Chemical Fire

On the evening of July 27, 1970, just after 6 p.m., a two alarm fire broke out at the Midstate Chemical Company located on the cities northeast side. The buildings warehouse held some seventy 150-pound cylinders of liquid chlorine which exploded creating a cloud of toxic fumes. In all, 10 firefighters were hospitalized and an additional 50 others were treated at the scene for smoke inhalation as this five hour fire forced the evacuation of 1200 plus residents of the eastside.

M.P.
612 223

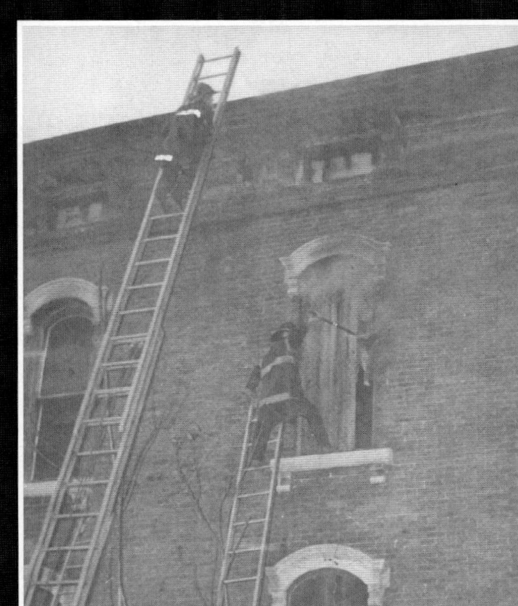

1 April 1971

Engine #14 ran out of Station #16, Ladder #14 out of Station #23, and Squad #14 out of Stations #21 and #22 until the new station opened on February 2, 1972.

1 July 1971

A new vacation schedule was adopted, firefighters got three weeks for 10 years of service, four weeks for 10-20 years of service, and five weeks for 20+ years of service.

1 January 1971

The department started paying holiday pay, ten paid holidays at $30.00 for each holiday worked.

The biggest change to ever happen to the run load of the IFD would be to start doing EMS, which began in the early 1970's. The responses and number of apparatus would change over the years. This was one of the biggest concepts for the members to buy in to and believe in. Most did so hesitantly, but by 1973 the department would be 100% trained in Advanced 1st Aid. Then in 1975 Chief Lamb dissolved first aid runs and the fire department was only used to back-up Wishard ambulances. This would drop first aid runs from 522 in May to 61 in June. This would be short lived as Chief William Patterson would restore first aid runs the following year as well as get six firefighters trained as Emergency Medical Technician (EMT's). The EMT program would be the future. In 1977 there would be 76 EMT's leading to 170 EMT's by 1978 giving the department enough to put at least one EMT per apparatus per shift. This would set-up full department responses to first aid runs in 1979. The department would make the EMT certification a requirement in 1980. This firmly planted the roots of the IFD doing first response EMS.

(See following text on page 104)

1 January 1971

The department started paying ride-out pay for non-officers and non-engineers performing these tasks while the regular firefighter was off duty.

25 May 1971

Engine #27 ran out of Station #12 and Ladder #27 out of Station #11 until the new station opened on March 3, 1972.

1971

All downtown companies got air horns put on their apparatuses to help move traffic.

1971

The 1st Black Expo was held.

1971

The new grooming policy was put into effect.

1971

The position of Deputy Chief of Operations and Administrative Services was created.

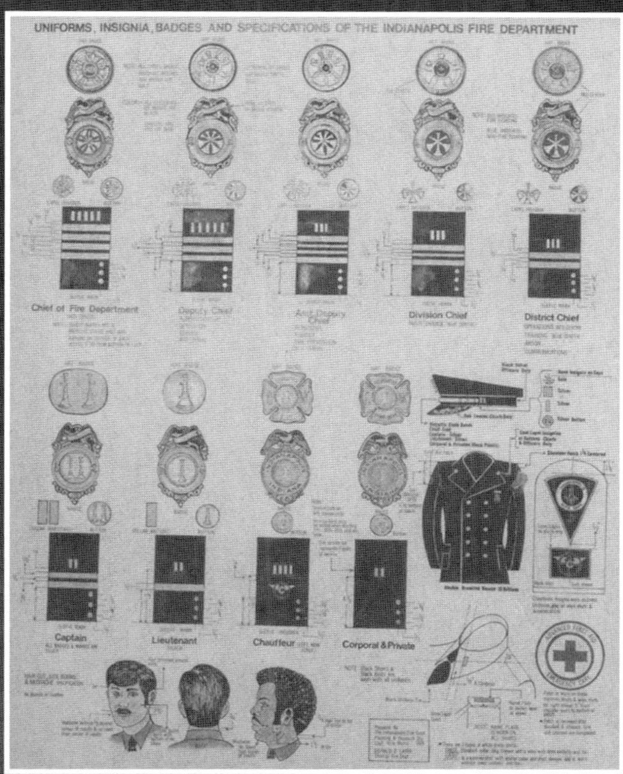

Richardson Rubber Fire

1 October 1972

The Centrex telephone system was established.

20 July 1972

Two rescue ropes were added to all engines.

5 April 1973

A Dictaphone and recording equipment was placed in service at dispatch to record all incoming and outgoing calls.

25 June 1973

Ladder #22 closed.

5 November 1973

The Grant Fire yielded $5,300,000.00 in damage using 33 pieces of equipment with fire damage to 15 buildings.

9 November 1973

Ladder #22 reopened.

The IFD in the 1970's seemingly was trying to find its identity. The department started purchasing unconventional equipment, some of it was brilliant, while a few were rather curious.

In October of 1972 they added a Centrex line to every firehouse. This is a dedicated phone line for department business only and dialed like an extension not a phone number. As a general rule, that phone line was to be used for serious business. Therefore, it was necessary to answer the line within two rings. This practice is still strong today within the firehouse.

Grant Fire

At 12:49 p.m. on November 5, 1973, Engine #13, Engine #7, Truck #7, and Car #2 were dispatched to the empty Grant Building on East Washington Street with the report of smoke coming from the building. Before the first unit could arrive, dispatch started receiving multiple calls for a building fire that would last several days to completely extinguish and even weeks for fire prevention and arson to investigate. In all, 220 fire fighters were used to fight this multiple alarm blaze on Box 45 bringing 18 Engines, 7 Aerial Trucks, 2 Snorkels, 1 Squirt, 4 District Chiefs, 3 Hose Tenders, 1 Light Wagon, 1 Division Chief, and a recruit class to the scene. 84 businesses in 15 buildings sustained damage as firefighters attempted to get a hold on a fire that would generate over $5.3 million dollars in damage.

Hofmeister Gym Fire

25 June 1973

District #2 was moved from Station #8 to Station #22.

1973

100% of the department completed training in advanced 1st aid training.

1973

A Chief's salary was $19,293.75 and a private's salary was $7,717.50.

1973

Chief Donald E. Lamb became the youngest hired chief to date at the age of 44.

1973

Department staffing consisted of 875 firefighters.

1973

Station #8 and Engine #8 closed.

Retired Indianapolis Firefighters Club

While attending a funeral of a retired firefighter in 1974, a suggestion was made to have a meal after upcoming funerals. This was the start of the club. Retired firefighter Charles P. Hill was the clubs first president in 1974. In all, there have been 16 presidents. The club has two social functions a month including a breakfast and luncheon. The club has allowed many retirees to keep in contact with each other while participating in many fire department related activities throughout the city. The local 416 Union has been a huge contributor to the retirees club over the years.

In 1975 every apparatus received its own handheld radio. The first Hurst tools were purchased in 1977 and put on Rescue 7, then Squad 2 and 14 in 1978. These tools are commonly referred to as the "Jaws of Life." Along with innovative tools, administration and policy changes were made. Due to the hippie and free love movement at the time, the department had to tighten up its grooming policy in 1971 not allowing long sideburns, long hair, and only neatly trimmed mustaches and sideburns. Some of these rules were already in place but needed some attention. The administrative staff was expanded adding a Planning and Research Division in 1974 and Special Operations in 1976. There would be more diversity in the IFD as it would hire its first female (merit) firefighter, Byronna Slaughter, on March 23, 1978. Ms. Slaughter would not be with the IFD long quitting after just a few years. It was rumored that many of the firefighters were not fond of having a woman on the job and ran her off.

Although not a division or a staff position, the department did add a "White Room" at the repair shops in 1976 as a place to do repairs to face pieces and SCBA's. This is important as the firefighters were using their SCBA's more and doing training with them. In 1974 the department used the smoke house to train firefighters on the ad-

6 May 1974

The Sports Arena at 50 North New Jersey Street caught fire.

8 August 1974

The Retired Firemen's Club was started.

11 December 1974

Engine #6 re-closed.

1974

A bonus day off from work was given for perfect attendance, which was later referred to as the perfect attendance (PA) day.

1974

A total of 18 American flags were purchased and replaced at stations.

1974

A total of 27 department vehicle accidents were recorded.

First Female Fire Fighter (Merit)

The first recorded female firefighter in the United States was an African-American. Held under slavery by a New York City firefighter, Molly Williams, worked on Oceanus Company No. 11 in the 1780's. The first female (Merit) firefighter hired by the Indianapolis Fire Department was also an African-American woman named Byronna Slaughter, who was sworn in on March 3, 1978. After a short career and reports of harassment and lack of acceptance, she chose to leave the department.

vantages of SCBA's and to get them comfortable wearing them. Although the IFD had SCBA's since the 40's, not a whole lot of firefighters used them. Some didn't work well and they were cumbersome, heavy, and to be honest, the culture was that real men don't wear them. This mentality was wrong: dead wrong. In the golden age of firefighting burning materials were wood or cotton. In more modern time materials are all kinds of plastics that can be extremely dangerous to inhale. The mentality was changing; it would just take a long time. The department put out general orders on wearing SCBA's but it was not enforced for some time.

(See following text on page 111)

1974
The budget allowed for the replacement of 106 pillows and 374 pillowcases in the firehouses.

1974
The department had 21 fire service contracts with businesses outside of the department's service area for fire coverage.

1974
The Planning and Research Division was created.

1974
The Smoke House was used for 1st time for firefighters to get acquainted with using Self-contained Breathing Apparatus (SCBA) packs.

1 May 1975
The Data Processing of Dispatched records system allowed the alarm office to track run statistics.

29 July 1975
All new firefighters were required to reside in Marion County to be employed.

1975
8.39% of all calls were false alarms.

1975
A Chief's salary was $22,188.71 and a private's salary was $10,793.00.

1975
A total of 10,109 fire hydrants were in service.

1975
A total of 525 fire alarm boxes were in service.

1975
A total of 84 people were arrested for arson resulting in 34 convictions.

1975
All apparatuses received hand held radios.

1975

Chief Donald E. Lamb had 1st Aid runs dissolved from the department in May. The department was used as back-up service to Wishard Hospital for 1st aid runs seeing a drop in runs from May to June – 522 to 61.

1975

Department staffing consisted of 863 firefighters.

1975

Station meals on average cost from $2.50 to $2.75 for both lunch and dinner.

1975

The department received the 911 system and silent alarm system.

1975

The Fill The Bell campaign raised $9,200.00.

1975

The Indianapolis Fire Department had 20 fire service contracts with various companies and township fire departments.

H.P. Wasson Fire on the Circle.

1 February 1976

The mandatory Physical Examination Program started.

1 April 1976

Engine #33 closed and reopened as Squirt #33.

1976

A 3200 p.s.i. compressor was placed into service at Station #14 to refill bottles.

1976

All engines got monitor nozzles mounted on them.

1976

All squad's and ladder trucks got K-12 saws.

1976

Department staffing consisted of 863 firefighters.

The Fellowship of Christian Firefighters

The Indianapolis Chapter of the fellowship of Christian Firefighters is part of an international not-for-profit, non-denominational, ministry to the fire service, which began in Denver, Colorado in 1977. The specific objective is to encourage one another (Heb. 10-15). They are composed of Christian Firefighters, active and retired, Clergy, Fire Buffs, family and friends. Their ministry reminds everyone that they are not alone in sorrow during trials and tribulations. They minister by giving away bibles and sympathy cards at Union churches, hospitals and fire department functions as well as supplying support to widows and families of injured or deceased firefighters.

The IFD added EMS, equipment, staff, and started changing its culture, but that wasn't all. The communication and dispatch system went through some updates. In 1975 the department went with the 911 system and silent alarm system, as well as hand held radios. The same year on May 1, the department was able to track all run stats using the Data Processing of Dispatches record system. This, just two years after starting to record all calls coming in to dispatch (April 5, 1973).

It wouldn't take very long for this to start paying dividends as one of the most famous fires in Indianapolis would occur just a few months later on November 5, 1973. The Grant Fire would do $5,300,000.00 worth of

(See following text on page 114)

1976

Robert Temple became the 1st African-American dispatcher.

1976

Squad #1 and #33 went into service.

1976

The Air Force gave the department a foam firefighting apparatus called SCAT #1.

1976

The department had their 1st six trained Emergency Medical Technicians (EMT).

1976

The department opened up the "White Room" at the repair shops to maintenance fresh air masks.

1976

The department went to Nomex and pants allowing no rubbers.

1976

The Indianapolis Fire Department Print shop opened.

1976

The Specials Operations Division was started.

1976

Three foot high numbers were placed on apparatuses making it easier for chiefs and chauffeurs to communicate on the fire ground.

1977

100 firefighters were promoted to the ranks of Captain, Lieutenant, and Chauffer.

1977

32% of the department was found to be overweight or obese.

1977

91.4% of firefighters voted to strike due to a lack of a contract. Mayor Hudnut averted a strike with only 25 minutes to spare.

1977

A total of 156 cans of auto wax were purchased for the waxing of the apparatus.

EMS Coordinators/ Directors

Robert Zickler	1980-1992
Greg Gates	1992-1994
Mike Schank	1995
Greg Harris	interim
James McNair	1995-2003
Steve Auch	2003-2007
Brian Sanford	2007-2008
Scott Isaacs	2008-Present

1977

All Division Chiefs, District Chiefs, their aides, and Squad's were equipped with Handie Talkie radios.

1977

Department staffing consisted of 869 firefighters.

1977

Silent watch was introduced to the firehouses.

1977

Squad #17 closed and reopened as Squad #29.

1977

Squad #2 went into service.

damage to a total of 15 buildings while using 33 pieces of equipment and 220 firefighters, including the current recruit class. Dispatch would start silent watch (1977), which would allow firefighters to not listen to the radio systems to tell them they had a run. The tones alone would notify them. The communication system would start using Computer Aided Dispatch in 1978, commonly called the CAD system.

February 2, 1976 the department would make a physical fitness examination required for all firefighters. As the department experimented and moved along, all progress almost abruptly came to a halt in 1977. The firefighters were just moments away from going on strike, which would have been the first time in our history. The department voted 91.4% to go on strike and mayor Hudnut averted it by 25 minutes to spare. The firefighters were sick of working without a contract, for lousy pay and no benefits.

1977

The 1st Hurst rescue extrication tool went into service on Rescue #7.

1977

The 1st Promotion Merit board was formed promoting five new Chiefs.

1977

The busiest engine was Engine #5 with 977 runs.

1977

The Indianapolis Fire Department Chaplains were to wear class "A" uniforms and carry the rank of District Chief for the 1st time.

1977

The Medical Clinic opens for firefighters to conduct physicals and rehab on the job injuries, Monday-Fridays 8-4:30pm.

1977

The pension payout was $4,535,178.89.

1977

There were 42 position vacancies in the department.

7 February 1978

Ladder #32 closed.

3 April 1978

Ladder #32 reopened.

October 1978

Squad #2 and Squad #14 got the "Jaws of Life".

If the 1970's were a time the IFD was trying to find an identity, then the 1980's could be a time defined by expansion. Although not everything expanded, the department did see some contraction as well. The IFD went from six battalions back to four in 1981. The department would also do away with the rank of Corporal through attrition in 1985. This was a rank of service, (12 years) not merit. The Division Chiefs' would lose their aides in 1981. Aside from that, there was much growth.

(See following text on page 120)

1978
A total of 170 firefighters were trained as Emergency Medical Technicians (EMT's), giving each apparatus per shift an EMT.

1978
All recruits got High-Rise rope rescue training.

1978
Byronna Slaughter became 1st female (Merit) and 1st African-American female hired by Indianapolis Fire Department.

1978
Department staffing consisted of 869 firefighters.

Arson Squad Came Under Fire Prevention

May 1981 – May 1982	William G. Earle
May 1982 – February 1983	Robert Liming
February 1983 – March 1984	Robert McClain
March 1984 – January 1987	Max Brenton (IPD)
January 1987 – June 1989	Jack Dugan
August 1989 – July 1992	Thom Carr (IPD)
August 1992 – September 1994	Mark Froelich
October 2001 – November 2009	Rick Lintner

1978
During a blizzard that crippled the city, many firefighters used personal 4-wheeled vehicles and snow mobiles to respond to EMS calls.

1978
SCAT #1 was given to the Indianapolis International Airport.

1978
Seven firefighters were replaced by 15 civilians as dispatchers at fire communications.

1978

The Affirmative Action Panel was formed to handle grievances.

1978

The Computer Aided Dispatch (CAD) system was placed into service.

1978

The Indianapolis Fire Department had 19 fire service contracts with various companies and township fire departments.

1 June 1979

Headquarters at 555 N. New Jersey Street was opened.

1979

Department staffing consisted of 863 firefighters.

1979

The department returned to making EMS runs.

1979

The Federal Emergency Management Agency (FEMA) was created.

1979

The Indianapolis Fire Department had 17 fire service contracts with various companies and township fire departments.

Show and Tell at Western Electric Plant.

The IFD started EMS a few years earlier and in typical IFD fashion wanted to branch out more and become more valuable to the citizens of Indianapolis. The IFD started Hazard Materials in 1984 making Station #13 a Hazard Materials house. The department would then add Station #12 to Hazard Materials house months later. Hazard Materials would not be the only expansion of services as the IFD would start a Self-Contained Underwater Breathing Apparatus (SCUBA) program in 1985; this was placed at Station #7 and referred to as the dive team. Then, two years later the department would add a second dive team and put it at Station #14. A year later in 1988 the department would expand yet again by adding a trench, collapse, and confined space team to Station #1 & #5. All of these teams were assigned TAC units that carry many specialty tools and equipment. They also undergo many hours of vigorous training. It is important to understand that while all these programs and disciplines were getting started with all the training that was required for these disciplines, the department continued training all firefighters to become EMT's. All of this was a pretty daunting task. As the department got more and more EMT's, apparatus were put in service every month as EMS apparatus. The last non-EMT would retire in 1997.

(See following text on page 125)

15 February 1980

The Arson Squad started co-staffing with Indianapolis Police Department.

18 June 1980

Recruit class #53 graduated the 1st class to participate in a mandatory physical fitness program.

1 August 1980

The average response time for incidents was 2:45 minutes.

1980

All firefighters had photo identification cards made.

1980

Department staffing consisted of 840 firefighters.

Hazmat Coordinators/ Directors

David Cutshaw	1984-1994
Tony Rossa	1994-2003
Tim Robinson	2004-2008
Fred Shoewmyer	2008-Present

Crystal Flash Fire

1980

Firefighters under the 1937 pension were offered a one-time buy-out payment of $10,000.00 to convert to the 1977 pension.

1980

In the 1980's, IFD had 73 2nd alarm or better fires, an average of 7.3 a year.

1980

Job descriptions for personnel were established.

1980

Light Wagon #13 went into service.

1980

The estimated Indianapolis population was 700,807 people.

1980

The oral interview became part of the interview process.

1 November 1981

The Fire Academy was moved to 901 North Post Road.

Special Operations Coordinators/Directors

Robert Zickler	Asst. Chris Pitts
Chris Pitts	Asst. Sony Ridout
Steve Dixon	Asst. John Scott
Mickey Radez	Asst. John Scott
Paul Sullivan	Asst. John Scott/Robert Albrich
Warren Todd	
Greg Gates	
Dave Owens	Asst. Trevor Hanshew
Trevor Hanshew	

1981

A new transfer/reassignment policy was established for firefighters wanting to move stations.

1981

Department staffing consisted of 817 firefighters.

1981

Station #14 sustained heavy damage from a kitchen fire.

1981

Station #9 closed and Engine #9 moved into Station #1.

1981

Tactical #7 went into service.

1981

The department went from six to four districts.

N. Pennsylvania Avenue.

Mutual Aid Responses

As the city of Indianapolis grew, so did the need for additional fire departments on the out-skirts of the coverage area of Center Township. Over time, eight additional Township Fire Departments, three city or town Fire Departments, and one airport Fire Department went into service in Marion County.

As early as 1978, the Indianapolis Fire Department had service contracts with these other fire departments and individual businesses outside the primary coverage area to provide "mutual aid" or fire service when the situation called for it.

Prior to Consolidation on January 1, 2007, Marion County, Indiana was serviced by the following Fire Departments:

Indianapolis Fire Department, Decatur Township Fire Department, Franklin Township Fire Department, Perry Township Fire Department, Pike Township Fire Department, Warren Township Fire Department, Washington Township Fire Department, Wayne Township Fire Department, Beech Grove Fire Department, City of Lawrence Fire Department, Speedway Fire Department, and Indianapolis International Airport Fire Department.

Experimental Squad's

In the late 80's, an experimental squad, which was a two man EMS van was placed in service. This van took on different call signs as it went from house-to-house to test the need for a squad at particular stations. It caught runs from Stations #5, #11, #22, and #31 during this experimental time. Currently none of these houses have squads.

Although not an expansion of service, the department would give engineers more respect by making a 1st Engineers Certification in 1983. They would then make a 2nd Engineers Certification in 1985. In the same year, the IFD would hold its first joint recruit class with Perry Township Fire Department which had 15 recruits. The engineer certification was very necessary as fire companies were dealt two game changers in 1988. The first being that firefighters were not allowed to ride on the tailboard of apparatus. This made responding to and coming back from runs much safer and would allow the engineer to drive faster more safely. The second was Large Diameter Hose (LDH), meaning 5-inch supply lines allowing engine companies to flow more water. As the fires grew in size, so did the tools that were used to fight them.

1981
The Division Chiefs lost their aides.

5 January 1982
Engine #9 ran out of Station #1 until February 17, 1984.

2 February 1982
Subs started paying house dues at the house they were assigned to during payday weekend.

15 March 1982
Ladder #32 re-closed.

1982
Department staffing consisted of 781 firefighters.

1982
Light Wagon #13 was moved to Station #7.

1982
The busiest engine was Engine #12 with 1531 runs and busiest truck was Truck #7 with 518 runs.

1982
The Fill The Bell campaign raised $5940.00.

1 May 1983
Engine #17 closed.

1 September 1983
A new state law was passed enabling widows to remarry without losing the pension benefits.

1983
Department staffing consisted of 774 firefighters.

1983

The 1st Engineer Certification Program was held.

1983

The department received computers in the Personnel Department.

17 February 1984

Ladder #17 closed and reopened as Ladder #4.

17 February 1984

Ladder #19 closed.

1984

A total of 69 department vehicle accidents were recorded.

1984

Department staffing consisted of 752 firefighters.

1984

Squad #41 went into service.

1984

The Hazardous Materials (HAZMAT) response team was established at Station #13.

1984

Station #12 became a Hazardous Materials Response Team, 2nd team.

1984

The Self-contained Underwater Breathing Apparatus (SCUBA) Dive Rescue team was established.

3 May 1985

The department did away with the rank of Corporal, which indicated 12 years of service on the department.

1985

1st Civilian Director of the Indianapolis Fire Department Communications Branch was hired.

Moma Ritchie's Worehouse Fire

1985
A total of 550 firefighters were trained as Emergency Medical Technicians (EMT's).

1985
A total of 73 Motorola HT400 handheld radios were purchased.

1985
Recruit class #59 was a combined class with the Perry Township Fire Department with 15 recruits from Perry.

1985
Recruit class lasted 13 weeks.

1985
The Confined Space Tactical Team was established and ran out of Stations #1 and #5.

1985
The second engineer certification program was established.

Terrace Street
and Madison Avenue

Redefining IFD 1986 - 2009

As the IFD has grown and expanded its services, the department felt compelled to start letting firefighters off work for some approved classes in 1987.

The IFD was moving in new directions and filling needs that were previously unnoticed. They started the Fire Stop Program in 1986, which counseled 130 children that were involved in 87 fires. The department found the need to take care of its own by starting a Critical Incident Stress Debriefing Team (CISD), which is something that firefighters used to adjust to the stresses of incidents that have a profound effect on one's emotional state. Another example of this new thinking would be that the IFD started a Safety Officer Program in 1989. This position tracks firefighter injuries as well as goes on all fires and tactical runs to help keep all personal working on the scene safe.

Adding programs and services weren't the only things going on in 1980's. The department would write and define all job descriptions. They also tried to buy all 1937 pensioners to the 1977 pension for a one-time offer of $10,000.00. A few years later in 1988 the department would form a Standard Operating Procedures (SOP's) committee to ensure that the focus of firefighting hasn't suffered with all the new and continued services now offered by the IFD. In conjunction

21 March 1986

Personal Fitness Evaluation (PFE) was established and used for the promotional process and became mandatory for all firefighters.

21 May 1986

The Fair Labor Standards Act (FLSA) took effect.

1986

Department had one of the fastest response time in the United States at 2.5 minutes per run.

1986

Department staffing consisted of 727 firefighters.

1986

Holiday pay was $55.00 per holiday worked.

1986

The Fire Stop Program was started and 130 children were counseled that were involved in 87 fires.

1986

The second dive team was added.

1986

The training tower was dedicated by Mayor Hudnut.

Market Street Parking Garage Fire

with keeping focused on firefighting, all the new disciplines, and EMS, the scope of the fire department was very wide. It was a time when many firefighters wanted to know what was going on or what was coming down the road so the department teamed up with a local cable provider to start the INET. This is a dedicated TV line that with a flip of a switch would turn to IFD designated programming. This was used quiet often for the Chief of the department to give updates and firefighters could call in and ask questions. This was also used as a training source for fire, EMS, and other disciplines. Also, awards banquets and ceremonies were shown on this channel for those whom were unable to attend.

(See following text on page 138)

7 July 1987

All firefighters were required to wear seat-belts while the apparatus is in motion.

1 August 1987

Squad #41 closed and reopened as Squad #13.

19 October 1987

Time off for certain department related classes was granted by the department.

First Black IFD Fire Chief

In 1955, black fire captain Rensil Williams retired from the Indianapolis Fire Department, making way for the hiring of black firefighter Joseph Kimbrew. After spending time at Station #1 and Station #9 in Haughville, Joe transferred to busy Station #8 after department integration. In 1968, Joe received the departments first "firefighter of the year" award. On January 19, 1987, Joe became the first, and only, black fire chief for the Indianapolis Fire Department.

23 October 1987

Smoking was only allowed in apparatus bays in the firehouses.

1987

Chief Joseph D. Kimbrew was 1st and only African-American Chief.

1987

The Advanced Emergency Medical Technician (AEMT) program was started.

Coke Company on Northwestern Avenue.

14 November 1988

The department hired the 1st female Deputy Fire Marshal.

1988

85% of all firefighters met the department's minimum fitness standards.

1988

A total of 13 firefighters were injured in the Murat Shrine Club Fire.

1988

All engines were equipped with 5" supply lines.

1988

Department staffing consisted of 756 fire-fighters.

1988

The 1st Physical Fitness Evaluations (PFE) were given.

Murat Shrine Temple Fire

On February 2, 1988, a fire broke out in the Murat Shrine Temple located just a couple of hundred yards southwest of IFD Headquarters. While fighting this multiple alarm fire, an overhead loft that was occupied by several firefighters, crashed down on additional firefighters below trapping several for up to 40 minutes. In all, thirteen firefighters were injured.

Fire at W. 30th Street.

14
STAY BACK
500 FEET

St. Michaels Fire

E. 21st Street Fire

1988

The city established a long term purchasing plan for apparatus.

1988

The city passed a smoke detector ordinance.

1988

The Critical Incidents Stress Debriefing (CISD) Team was created.

1988

The Decontamination Team (Decon) #19 was formed with Engine #19 and Rescue #19.

1988

The department started a five-year plan to put 5" hoses on all engines.

1988

The department stopped allowing firefighters to ride on the tailboard of apparatus.

Heading into the 1990's the IFD was riding high and even started to upgrade its communications system. By 1990 all the dispatchers were civilian, saving the city money. The enhanced 911 system was tested in 1989 and went into service in 1990. In 1991 MECA became fully functional and went digital a year later. MECA is the hub of dispatching for public service for the whole county. This wasn't always the case as many smaller fire departments would have their own dispatching system. They would soon want the new technology, better staffed, and more experienced MECA dispatching. The different dispatching systems caused quite a bit of trouble when mutual aid was given to departments on runs, but by 1993 seven township fire departments would be using MECA.

1988

The Incident Command (IC) System was put into place.

1988

The position of Battalion Chief of Hazardous Materials was created.

1988

The safety office was designated to investigate all firefighters injuries and make safety suggestions to firefighters.

1988

The Standard Operation Procedures (SOP) manual committee was formed.

1988

The Trench Rescue discipline was added to the Confined Space Tactical Team.

1989

The city hosted the 116th International Association of Fire Chiefs.

1989

The Internal Network (INET) was being used for Emergency Medical Services (EMS) refreshers classes as well other trainings and department-wide communications.

The IFD proudly forged ahead in the 1990's with a seeming air of invincibility. This would come crashing down as the IFD would be shaken to its core by a fire that would make national news for two reasons; one it claimed the lives of two firefighters, and two it happened where the Mike Tyson jurors was sequestered. In the early hours of February 5, 1992 the IFD responded to the Indianapolis Athletic Club to investigate an alarm. They would find smoke on three floors. After they thought to have the fire knocked down, a false ceiling would come down where a majority of the fire was growing. This instantly changed the conditions for the crews operating inside causing many crews to bailout and some to become disoriented. Two firefighters would make the ultimate sacrifice and many others would be injured. This one incident would be one of the most important fires in the history of the Indianapolis Fire Department. In the months and

Indianapolis Athletic Club Fire

At 12:06 a.m. on February 5, 1992, a box alarm was dispatched to the Indianapolis Athletic Club located at the corner of Meridian Street and Vermont Street for a smell of smoke. Truck #7 arrived minutes later and reported "on the scene, nothing showing, will investigate." This fire alarm at 350 North Meridian Street was anything but routine. While looking for the source of the smoke, the alarm panel showed alarms going off in the basement. The fire was found 11 minutes later on the third floor. While fighting the fire, a flashover occurred killing Corporal Ellwood Gelenius and Private John Lorenzano. The fire killed an additional civilian as well as injuring four other firefighters. From this fire, the HOG manual, High-Rise Fire Ground Manual, was credited to establish accountability for fire companies while on high rise fires. This fire made national attention, not only from the firefighters that perished, but because it was the hotel where twelve jurors and thirteen alternate jurors for the Mike Tyson trial were staying.

19

years to come, the members of the IFD would continue to ask themselves questions like, Did we do something wrong? Did we over look something? Could we have done something different? What can we do to keep this from happening again? All of these questions are difficult to answer. The IFD took a good long look in the mirror and decided to do more. They decided not to let two brothers die in vain, and to not forget!

The IFD response would be swift and long lasting. The IFD would create a High-Rise Operations Guide (HOG) that would get national attention. Along with a high-rise chief position that responded on all high-rise fires and alarms, a new kind of high-rise training for the department was created. Members found themselves rejuvenated to the job and to carry on John and Woody's memory. They were doing all the little things that maybe weren't getting done before. The IFD bought a flashover chamber in 1994, which was one of only a couple in the country. This would teach firefighters the signs of one of the most dangerous fire conditions, a flashover. A flashover occurred at the Athletic Club, which was a big contributing factor to the death of John and Woody. Recognizing fire conditions is a good start but not enough. The department would start doing Rapid Intervention

Indianapolis Black Firefighters Association

On October 22, 1992, the Indianapolis Black Firefighters Association (IBFA) was established in the basement of Mt. Zion Church at 3435 Boulevard Avenue with their goal to lend a helping hand: through education, training, and providing essential resources to firefighter, our youth and community, the IBFA finds its current headquarters at old Station #31 located at 1201 East 46th Street. The IBFA sponsors several community programs including Christmas at the firehouse, a Halloween haunted house, a neighbor picnic, bikes in the hood – member ride their bikes in the neighborhood handing out smoke detectors, and they also prepare breakfast for IFD retirees once a month at the retiree's fellowship breakfast.

1989

The Safety Officer program was started.

1989

The total department budget was $29,346,259.00.

11 July 1990

The Concerned Firefighter Association of Indianapolis was established. In 1991 it was renamed the Indianapolis African-American Firefighters Association.

1990

A total of 70 firefighters ran in the Indianapolis Mini Marathon.

1990

The communications division, except for the director, was staffed by all civilians.

1990

The enhanced 911 system that was tested in 1989 was placed into service.

1990

The estimated Indianapolis population was 731,327 people.

21 February 1991

Squirt #33 closed and Engine #33 reopened.

1991

A smoke detector law went into effect requiring all one and two family dwellings must have at least one working smoke detector.

1991

Department staffing consisted of 756 firefighters.

Team (RIT) training in 1997. This is quite simply a way for firefighters to save firefighters. Many cities had started this but most were unsure of exact tactics to extract firefighters. The IFD was no different. Over the years protocols would change as well as equipment needed to perform this task.

The IFD would become NFPA compliant, which would open a lot of doors for funding and equipment. The department had at least one handheld radio per apparatus and pushed to get one for every riding position. The IFD would come to realize that its fire gear was outdated. This led to the purchase of new and better gear followed by the purchase of a second set of gear. This new fire gear would be tan and give the IFD a new look. Along with this new look, uniforms would change as well. The department decided to go with fire retardant station wear. These new fire retardant cloths would be expensive costing around $40.00 per tee shirt. All the new purchases would mean the department would add another position in 1992 known as Materials Management. A year later it would be known as the quartermaster. This position would order and exchange fire gear, cloths, and equipment. The age and condition of fire gear and station wear was something noticed after the Athletic Club. A larger and more expensive purchase was made as well with upgrades that would assist in firefighter safety. The department would buy new SCBA's going away from Scott brand and buying Interspiro. These new air packs had an integrated PASS device to assist firefighters in finding other firefighters in trouble. This is a device that

Wishard Ambulance Garage Fire at 4040 N. Rural Street.

Rapid Intervention Team

In late 1997, the Rapid Intervention Team (RIT) procedures were established. The RIT team was designated as a team of firefighters that would be dispatched to working fires with the sole purpose of rescuing trapped, lost, or injured firefighters on the fire ground. The battalion chief's vehicle had a yellow bag with such tools as bolt cutters, rope, door chalks, and others that could be utilized while extricating a firefighter. As the importance to having this safety team on the fire ground grew, so did the equipment upgrades and training scenarios for the firefighters. In 1999 a make-shift air pack was added to the bag. With the change to Scott's SCBA in 2002, a new RIT bottle was upgraded. Two major trainings took place in 2001 and 2006 that have placed firefighters in interesting scenarios to think outside the box if they become trapped, lost or injured.

The dispatching order went from sending a RIT apparatus once companies on-scene marked the incident a working fire to a more proactive approach of dispatching an engine company with the initial companies on any report of fire. The engine company that is designated as RIT is based on its order of arrival on the scene, usually the third engine on a still alarm, the fourth engine on a box alarm, or the fifth on a high rise box alarm. An additional ladder company is dispatched as the working RIT company when the incident is marked a working fire. This second RIT company performs proactive tasks such as additional laddering, forcible entry or ventilation, and as a relief RIT if the initial RIT is deployed. If a designated RIT is moved up to assist with fire attack, an additional RIT is dispatched to fill its spot.

Once activated, each RIT mayday brings and additional RIT task force which consists of an additional battalion chief, engine, ladder, and transporting medic unit. Currently, all ladder companies, TSU's, battalion chiefs, and additional support vehicles have a RIT bag on their apparatus making it easier for a RIT company to deploy regardless of an apparatus's arrival to the fire ground. Since introducing the current SOP in 2007, the word "mayday" has been used nine times resulting in the deployment of the RIT team four times. All nine incidents resulted in zero firefighter injuries.

1991

The Metropolitan Emergency Communication Agency (MECA) Dispatching system was fully functional and running.

5 January 1992

The Personal Alert Safety System (PASS) devises were added to Self-contained Breathing Apparatus (SCBA) packs.

10 January 1992

The 1st day for digital dispatch.

1 June 1992

Engine#4 and Ladder #4 ran out of Station #19 until March 10, 1996 when Engine #4 was closed and Ladder #4 was renumbered Ladder #19.

1992

All front-line apparatus had a diesel exhaust filter system installed to eliminate harmful fumes.

1992

Department staffing consisted of 746 firefighters.

1992

Station #31 became a Hazardous Materials (HAZMAT) Station.

1992

The 1st annual awards banquet and dinner was held.

Tanker Rollover

when left sitting still for about 30 seconds will activate with a chirp then gets louder and louder until going into a full alarm mode. It allows firefighters to follow the noise to a downed or missing firefighter.

Adding a quartermaster and a high-rise chief weren't the only administrative changes of the time. A new records system was put into place called Record Management Systems (RMS). This unified the reports and allowed for easier recall and statistical data. This went into place in 1993, the same year the department would move its repair shops to south Belmont Street with other city vehicles. 1993 would prove to be a busy year as the committees for the new SOP's were completed and put in place. This was also the year that the IFD would start co-staffing with Wishard ambulances. The substitute firefighters would work 12 hours on an ambulance then go to an assigned firehouse for the other 12 hours. This would last until 2000 with a very welcomed termination of the program.

1992

The Training Division and the Special Operations Division were combined to form the Education Division.

1992

Wishard Hospital joined the Metropolitan Emergency Communication Agency (MECA) Dispatching system.

1 January 1993

Sick leave went from accrued time to 90-day sick leave for new hires.

1 January 1993

Special position pay (stack pay) started for Rope, Dive, Hazard Material (HAZMAT), Confined Space/Building Collapse/Trench (CRT), Emergency Medical Technician (EMT), Advanced Emergency Medical Technician (EMT), House Captain, Dive, and Urban Search Team.

The Saint Florian Center
Sparking the fire in Indy's Youth

St. Florian Center

Developed by Indianapolis firefighters in 1992, the St. Florian Center's mission is to provide at-risk Indianapolis youth an opportunity to foster leadership skills, develop problem solving methods, and survival tactics through a variety of programming and opportunities in the community in order to create leaders of tomorrow. Youth are influenced to "do something positive – be someone positive – have something positive". With various programs, the St. Florian Center has reached out to thousands of youths in the Indianapolis area.

www.saintfloriancenter.org
2511 East 46th Street Suite P-1
Indianapolis, Indiana 46205

Fire at Terrece and Madison Avenues.

Coming off the heels of a double firefighter fatality, the department and the city started seeing all the trials and tribulations that firefighters go through and started an awards banquet in 1992 to show appreciation. This is a nice "thank you" to a humble department. It is hard to get an answer out of a firefighter other than "I was just doing my job" when it comes to getting recognized for a job well done.

In 1995 the city of Indianapolis would get a contract to host the Fire Department Instructors Conference (FDIC). This is one of the largest conferences for the city and was the perfect place to display the talents of the IFD to the public and the firefighters from all over the world. The IFD has long been at the forefront of the fire service and when put on the world stage for our profession, we have shined. Firefighters from around the world have witnessed the midwestern work ethic mixed with new tactics and aggressive fire attack. The IFD of the past and present have had a hand in shaping the fire service of the future.

(See following text on page 155)

1 January 1993

The Bereavement Leave Program was established for firefighters who lost loved ones with firefighters receiving two 24-hour shifts off for the death of a parent, spouse, or a child.

1 April 1993

The department returned to BOX dispatching.

1 August 1993

The Indianapolis Fire Department and Wishard Hospital ambulance co-staffing started at Stations #1, #12, #29, and #33.

A City of Hosting

The City of Indianapolis has had the honor of hosting several fire department related conferences and events over the years including the following three.

The annual Fire Department Instructors Conference (FDIC) moved to Indianapolis in 1995 and is currently still here. FDIC brings in close to 30,000 firefighters from through-out the world for hands-on and class room trainings.

The biennial event of the World Police and Fire Games that started in 1985 was in Indianapolis in 2001 with over 6500 participants competing for medals. Other cities to host the games include:

1985 San Jose, California
1987 San Diego, California
1989 Vancouver, British Columbia, Canada
1991 Memphis, Tennessee
1993 Colorado Springs, Colorado
1995 Melbourne, Australia
1997 Calgary, Alberta, Canada
1999 Stockholm, Sweden
2003 Nevada, Las Vegas
2005 Quebec City, Quebec, Canada
2007 Adelaide, Australia
2009 Vancouver, British Columbia, Canada

And in July of 2005, Indianapolis hosted the Indiana Fire Chiefs Convention.

1993

A Chief's salary was $49,143.00 and a private's salary was $33,059.00.

1993

A females quarters was built at Station #2.

1993

An accelerant-sniffing dog named "Knight" joined the Indianapolis Fire Department Arson Division.

1993

Department staffing consisted of 723 firefighters.

1993

Firefighters were given free parking to visit the chief at the City County building expressing his open-door policy to firefighters.

1993

Holiday pay went from $80.00 to $100.00 per holiday worked.

1993

Rubber fire boots costs $65.87 a pair while leather boots cost $131.18.

1993

Seven township fire departments joined the Metropolitan Emergency Communication Agency (MECA) Dispatching system.

1993

Squad #1, #29, and #33 closed.

1993

The Basic Emergency Response Training (BERT) Program was adopted as the minimum training requirements for all Tactical Rescue Personal.

1993

The Belmont Maintenance Garage opened.

1993

The city agreed to pay a college incentive to firefighter at $200.00 per year finished at a maximum of $800.00.

1993
The department expanded its duel response areas with Washington, Warren, Pike, and Perry Townships.

1993
The department had bottle fill Stations at Station #7, #14, #29, and at the Belmont facilities.

1993
The department implemented the Records Management System (RMS).

1993
The department recorded 183 firefighter injuries.

1993
The department started a performance appraisals and job description standards.

Indiana Task Force 1 (IN-TF1)

The Indiana Task Force 1 (IN-TF1) was created in the early 1990's to respond to a variety of emergencies and disasters such as earthquakes, tornados, floods, typhoons, technological accidents, terrorist activities, hazard material releases, or hurricanes. IN-TF1 is one of 28 task forces that have been created under the National Department of Homeland Security (FEMA), National Urban Search & Rescue Response System. With IFD being the sponsoring department for this task force, highly trained personal in fields such as search specialists, heavy rescue specialists, hazard materials specialists, medical specialists, heavy riggers, structural engineers, and logisticians, have responded to several incidents including the Tippecanoe County Tornado (1994), the World Trade Center (2001), the Space Shuttle Incident (2003), the Evansville Tornado (2005), and Hurricane Katrina (2005). A type "I" task force, 80 personal, or a type "III" task force, 34 personal, must be able to deploy by ground within four hours and within six hours by air and can remain self-sufficient for up to 72 hours without restocking with a deployment lasting up to ten days.

FEMA

M & I Bank Building Rope Rescue.

N. Drexel Avenue

1400 Block of E. Lexington Avenue.

Campbell Avenue Smoke Explosion

1993
The department started hiring civilians for code enforcement.

1993
The K-Tool was introduced for forcible entry.

1993
The Material Management Department was started. It was later renamed the Quarter Masters System in 1993.

1993
The operating budget was $41,217,754.00.

1993
The safety response team was established consisting of Car #102, Car #104, and Car #742 for firefighters injuries.

1993
The shop mechanics became employees of the Indianapolis Fleet Services

In 1995 the department would experiment with a new position called the EMS Duty Officer (EDO). This was put in place for one year to see if it had an impact, and it did. So in 1996 it became a permanent position for the department. This position is responsible for getting EMS supplies to all front line apparatus and assisting with EMS and victims assistance at fires. 1996 is also the year the department received Kelly Days. The department had them in the past but as schedules change they weren't able to keep them. This dropped the working hours from 56 to 48 hours per week. Firefighters were able to pick one day per week as their Kelly Day and every three weeks when that day came around as their normal shift day, they would get it off. The firefighters often refer to the best Kelly day to have as the one that's coming up.

(See following text on page 158)

1993

The Standard Operating Procedures (SOP) manual was created.

31 January 1994

Battalion Chief #12 was moved from Station #20 to Station #25.

31 January 1994

Engine #12 closed and reopened as Squad #12.

31 January 1994

Ladder #18 closed and reopened as Strike Rig #18.

31 January 1994
Ladder #5 closed and reopens as Strike Rig #5.

31 January 1994
Light Wagon's #1 and #21 were renamed Tactical Support Units (TSU) #1 and #21.

31 January 1994
Squad #2 closed and Ambulance #2 moved into Station #2.

28 February 1994
Ladder #25 closed and reopened as Ladder #20.

22 June 1994
A Cool Down Program allowed hydrants to be opened to cool down people when temp was 85+ degrees Monday-Saturday, 1:00pm -6:00 pm.

31 January 1994
Ladder #34 closed and reopened as Ladder #2.

The Re-Organization Effort of 1994

Implemented in January, 1994, the department recognized a need to re-align its personnel and resources to better serve the city. Under the reorganization, the following changes occurred in Phase 1:

- Ladder #5 and Ladder #18 became Strike Rig #5 and Strike Rig #18, reducing the total staffing from 24 firefighters to 12 firefighters.
- Engine #12 became Medic Squad #12, reducing the total staffing from 12 firefighters to six firefighters.
- Ladder #34 was moved to Station #2 and Rescue #2 was replaced by Ambulance #34.
- Ambulance #5 moved to Station #5.
- Ladder #25 was moved to Station #20 and renumbered Ladder #20.
- Battalion #2 was moved from Station #20 to Station #25.
- Light Wagon #21 was moved to Station #7 and Light Wagon #1 and #7 became Tactical Support Units (TSU) #1 and #7.

The following changes occurred during Phase 2 implemented in the fall of 1994:

- Engine #28 became Medic Squad #31.
- TSU #1 and #7 increased staffing to three firefighters per unit.
- Rescues #7, #13, and #14 became advanced life saving (ALS) Squads.

The labor/management committee continued to evaluate the moves over the next two years.

E. New York Street

14 March 1994
Engine #32 ran out of Station #16 until June 15, 1994 while the station was being refurbished.

1994
Department staffing consisted of 726 firefighters.

1994
Survive Alive opened.

1994
The city agreed to allow firefighters to roll up to 10 days of vacation into the bank to be used at a later date.

1994
The department purchased a Sweden Survival Flashover Simulator for training.

The IFD entered the new Millennium organized, experienced, well staffed, and disciplined. The IFD was a well-oiled machine. The IFD would be the host department to the 2001 World Police and Fire Games. This is the third largest amateur event in the world, only behind the Olympics and the Pan Am Games, which Indianapolis hosted in 1986. The IFD would enter many teams as well as other area fire departments and win numerous medals. Once again the city and the IFD were on a world stage and handled ourselves with class, diplomacy, and a quiet confidence. This would not be the only new event hosted by the IFD as Indianapolis hosted the Indiana Fire Chiefs Conference in 2005.

Survive Alive

In the mid 80's, the firefighters of the Local 416 Union were doing an urban renewal project in downtown Indianapolis when they came across a Survive Alive House in Mishawaka, Indiana. After purchasing the old Fire Station #2 located at 748 Massachusetts Avenue in 1984, the Local 416 had a new home, and place to educate. Through the help of the memberships hands' on renovations, the Indianapolis Survive Alive House and Local 416 Museum opened in 1995. The Firefighters Museum and Survive Alive Inc., a non-profit organization, has educated over 200,000 people, mostly children, in their survive alive and shelter-in-place programs. Children are taught about smoke detectors use, fire drills, stop-drop-and-roll, and fire safety in the house. The Indianapolis Metropolitan Police Department also gets to educate the children about life safety issues including gun safety, motor vehicle safety, stranger danger and other life safety techniques. Firefighters Survive Alive is open to the public, by appointment, daily from 8:00 a.m. - 4:30 p.m.

Firefighters Survive Alive
748 Massachusetts Avenue
Indianapolis, Indiana 46204
(317) 327-6707

FIREFIGHTERS SURVIVE ALIVE!
SERVING INDIANAPOLIS AND MARION COUNTY

White River Water Rescue

In 2004 the departments' contract with its fire gear would expire and would start a bidding and testing committee. The department would settle on Globe brand and go back to black gear. The first set would take a long time to get out to the firefighters because of the military contracts that Globe had to fulfill before making fire gear for the public sector. Also the department would start testing new SCBA's and after field and noncombat testing they would go back with Scott brand and a more classic IFD look. The department went back to a traditional look but a new age mentality adding an Accountability Officer in 2007, a Shift Training Officer in 2008, and offering a second officer class. Any private wanting to ride out in the officer seat was required to attend this class.

1994

The department went through a large reorganization reducing the number of apparatus on the department.

1994

The telecommunications Company Cellular One paid for renovations to Station #32 in return for a cell tower and contract on the property, which is still in use today.

Downtown I65 accident involving two buses and a truck.

2005 would be a pivotal year for the department first implementing a Work Performance Evaluation (WPE), one of two cities nationwide to do so. This is an incumbent physical agility course designed to make sure all members are fit enough to perform all tasks on a fire. As well as making the department a tobacco-free workplace. There was to be no more smoking on city property. The department would provide assistance to firefighters who wanted to stop their tobacco use by providing nicotine gum, patches or other necessary materials. The department seeing change on the horizon would edit, train, and distribute new Fire Ground Operations Guides (FOG's) to reinforce tactics, decision making, discipline with fire ground assignments, and accountability.

A few years prior, the city offered the department a "drop plan" meaning members who were eligible for retirement could sign a contract for 1,2 or 3 years freezing their pension rate and receiving a lump sum amount at the rate their pension was froze at for the duration they choose 1,2,or 3 years at retirement. Although some members had retired before using this in 2005, 69 firefighters would retire accounting for 2469 years of experience. This would account for almost 10% of the department and most senior members including a lot of officers and many chiefs. The department would go through a large promotion process, promoting around 10 chiefs, 18 captains, and close to 40 lieutenants. This would come on the verge of one of the biggest things to happen to the IFD; consolidation.

The Fraternal Order of Leatherheads Society (Fools - Indianapolis Charter)

In 1995 the Fools Organization was started by a group of firemen in Central Florida. This group had a passion for camaraderie and brotherhood that is forged between firefighters. The term "Leatherhead" refers to the leather helmets that are worn by so many firefighters throughout the country. This society brings together firefighters to share stories, ideas, and training techniques to make them all better firefighters. The Indy Metro Fools Chapter was started in August of 2002.

16th Street and Central Avenue

19 March 1995

Engine #28 closed.

1995

Department staffing consisted of 722 firefighters.

1995

Station meals on average cost from $5.00 to $6.00 for both lunch and dinner.

1995

The city agreed to provide parody with the Indianapolis Police Department when negotiating union contracts.

1995

The Emergency Medical Services Duty Officer (EDO) position was created for a 1-year trial period.

1995

The Fire Department Instructors Conference (FDIC) was moved to Indianapolis.

1995

The house captain received $50 per staffing position for houses that met the new cleaning standards.

1995

The Indianapolis Fire Department was the 22nd busiest fire force in United States, (*Firehouse Magazine* June '96).

In 1970 the passing of UNIGOV would lay the foundation for the mayor to push for consolidation of services with the police and sheriff and the IFD and at least eight township fire departments. The mayor would call this plan "Indy Works". It would come as a complete surprise to the rank and file of the IFD and other township fire departments. The union and many members from all potentially affected departments partnered up to form committees to compile framework for an as painless as possible transition. Everything from uniforms, apparatus, responses, station locations, personnel, pay, and more were on the table to compare, agree upon, and find savings within. Although this would fail a couple of years in the legislature as a full consolidation, one department would show interest and eventually merge with the IFD.

(See following text on page 169)

Consolidation

During the first 150+ years, the department has grown significantly in numbers and size to match the growth of Indianapolis. In addition to building new stations and hiring new firefighters, the department has also grown by acquiring existing fire companies as well as consolidating with some others.

The first was on April 28, 1897 when the department annexed the West Indianapolis Fire Department. This department brought on an additional fire house and hose company to the west side of Indianapolis.

On June 20, 1922, the town of Broad Ripple was annexed by the city of Indianapolis. At the time, one voluntary station existed, which would be renamed IFD station #32.

In 1970, the consolidation of city and county governments, known as UNIGOV, paved the way for the consolidation of fire departments within the county with the approval of all parties involved. The bill referred to as "Indy Works" was introduced in 2005.

On January 1, 2007, the Indianapolis Fire Department (IFD) and the Washington Township Fire Department (Washington) consolidated adding 151 new firefighters and five new stations. Washington was started in 1941 as a bucket brigade by the name of the 6th District Volunteer Fire Department (Washington Township Outside) Inc. Located in the north side town of Ravenswood, it wasn't until August of 1848 that the name was changed to the Washington Township Volunteer Fire Department, Inc.

The relationship between IFD and Washington started as early as 1942 when the Washington's first engine was sent to the IFD shops to remove chemical tanks and add two water tanks giving it a capacity of 335 gallons. Additional programs such as service agreements, later referred to as mutual aid, and shared training recruit classes existed prior to consolidation.

Washington was able to obtain first aid training from the American Red Cross for all non-medical professionals in 1952. They went on to expand their EMS division with the introduction of their first ambulance service in September of 1964. In 1977, the department started classes at Butler University for their first group of paramedics. At the time of consolidation, Washington's services included EMS, suppression, extrication, hazmat, rope, and dive.

On July 1, 2007, the IFD and the Warren Township Fire Department (Warren) consolidated adding 115 new firefighters and five new stations. The Pioneer Volunteer Fire Department was started in 1937 and serviced the far-east side of the county. It later formed the Warren Township Civil Defense Volunteer Fire Department which was reorganized to the Warren Township Volunteer Fire Department in 1942.

Like Washington, Warren shared some similar services with IFD. Warren's first fire engine was received from IFD's reserve equipment. IFD and Warren also shared recruit classes and had shared mutual aid. At the time of consolidation, Warren's services included EMS, suppression, extrication, and hazmat.

On August 1, 2009, the Indianapolis Fire Department and the Perry Township Fire Department (Perry) consolidated adding 106 new firefighters and four new stations. Perry was started in 1950 and supplied fire service to the south side of the county.

Perry started with volunteers and paid chauffer making it a combination department. Their first paid firefighters were retired IFD and army personnel. Years later they would become a completely paid fire department. IFD and Perry also shared recruit classes and had shared mutual aid. At the time of consolidation, Perry's services included EMS, suppression, extrication, hazmat, and dive.

The growth of department continues with more mergers being discussed. Currently the department operates with 38 fire stations with the following: 37 engines companies, 20 ladders companies, six rescue squads, seven battalion chief's buggies, eight medic trucks, four tactical support units, 10 tactical units: 1, 5, 7, 13, 14, 19, 29, 31, 43, 44, three safety cars, two EMS duty officers, one accountability officer, one shift commander, and one public information officer.

2800 N. Sherman Drive

Haz Mat incident at State House.

1 February 1996

The position of Communications Director became civilianized.

10 March 1996

Ladder #4 closed and reopened as Ladder #19.

1996

A Chief's salary was $52,717.00 and a private's salary was $36,157.00.

1996

Department staffing consisted of 736 firefighters.

1996

Station #12, Engine #4 and #34, Squad #12, and #31 closed.

1996

The department started the "random drug test" program.

1996

The Emergency Duty Officer (EDO) was made permanent position.

Indiana Historical Marker

Erected in 1999 and located at 2 West Washington Street stands one of Indiana's historical markers which honors the Indianapolis Fire Department. It also recognizes the efforts of 86 firefighters who fought the Bowen-Merrill Fire on March 17, 1890, 13 who perished.

43

7 February 1997
Engine #21 closed and reopened as Squad #10.

1997
Started training RIT operations for the department

1997
The Catastrophic Event Planning Action Plan was created.

1997
The remaining seven non-EMT firefighters got trained as Emergency Medical Technicians (EMT's) or retired from the department.

1998
The Mobile Data Terminal (MDT), an onboard computer for run location and information, began being installed in front-line apparatus.

1998
The Survive Alive Car Seat Program was started.

1 January 1999
The firefighter Bill of Rights was established.

1 January 2000
A meal allowance of $400.00 per year was given to firefighters.

2000
All squads became Advanced Life Saving (ALS) Medic Squads.

2000
Engine #14 ran out of Station #5, Ladder #14 out of Station #31, and Squad #14 out of Station #16 until the new station opened October 2001.

2000
Tactical #29 Urban Search and Rescue was started.

The Washington Township Fire Department would be the first to merge on January 1, 2007. They would bring five stations, five engines, three ladders, four ambulances, one battalion chief, one TSU, and some tactical support for hazard materials, rope rescue, and top water rescue, as well as 151 firefighters. This transition went without incident but left members a little standoffish between the two formerly separate departments. The administration would take notice and when another department merged, the department would add steps to make it a more seamless transition. It wouldn't take long for this to happen either, as the Warren Township Fire Department would merge next just six months later on July 1, 2007. They would bring five stations, five engines, one ladder, three ambulances, one battalion chief, one hazard materials unit, and 115 firefighters. The Perry Township Fire Department would merge on August 1, 2009 and would bring four stations, four engines, two ladders, one battalion chief, and 106 firefighters. The final merger with Perry would create a need for a second safety officer located on the south side. As time passes, the belief is that additional departments will consolidate and possibly lead to total countywide consolidation.

(See following text on page 172)

2000
The Candidates Physical Aptitude Test (CPAT) was adopted as a stage for the hiring process.

2000
The department established a Post Incident Review Panel.

2000
The Emergency Duty Officer (EDO) was added to all working fires.

2000
The estimated Indianapolis population was 781,926 people.

2000
The Indianapolis Fire Department and Wishard Hospital ambulance co-staffing ended.

2000
The Indianapolis Fire Department taught 1073 civilian Cardiopulmonary Resuscitation (CPR).

2001
Station #14 became the only station to be rebuilt three times at the same location.

2001
The World Police and Fire Games were held in Indianapolis.

30 November 2002
Squad #29 reopened, 2nd time.

2002
The Indianapolis Fire Department and Washington Township Fire Department had a joint hiring process.

Cosmopolitan On The Canal Fire

Around 3:20 a.m. on March 12, 2009, a fire broke out at the corner of Michigan Street and Senate Street at the Cosmopolitan on the Canal. Over 100 firefighters fought this three alarm fire that housed a complex of retail and residential space that was under construction. The reported $28 million dollar project was ruled a complete loss from this arson ruled fire that took several hours to contain.

TACTICAL SUPPORT
6
INDIANAPOLIS FIRE
DEPARTMENT

2002
The physical abilities test was established for incumbent firefighters.

1 January 2003
The city doubled the life insurance amount for active firefighters to $50,000.00.

2003
A Chief's salary was $60,442.00 and a private's salary was $43,876.00.

1 July 2005
The city hosted the Indiana Fire Chiefs Convention.

5 December 2005
Squad #29 re-closed, 2nd time.

5 December 2005
Strike Rig #18 closed.

5 December 2005
Strike Rig #5 closed.

31 December 2005
Engine #32 closed and reopened as Squad #32.

2005
A general order for Vehicular Maintenance Responsibilities for apparatus engineers was created.

2005
A total of 69 firefighters retired totaling 2,469 years of combined experience .

2005
The department initiated the Work Performance Evaluation (WPE) Program, one of only two cities in United States.

2006
A total of 4,236 smoke detectors were distributed to Indianapolis residents.

1 January 2007
Former Washington Township Engines were renumbered; #221 became #21, #222 became #12, #223 became #6, #224 became #4, and #225 became #8.

From humble beginnings, a town became a city and a necessity became a passionate profession. The 150-year history of the Indianapolis Fire Department has been rich in mystique and character. At the Indianapolis Fire Department, we demand more of ourselves because our citizens deserve it. Throughout our history we have faced trials and tribulations head-on with unwavering resolve. This is not a safe job. We wake up in the morning and go to work not knowing for sure if we will come home the next morning. We pray we don't have to make the ultimate sacrifice, but the reality is, we have to take chances and risks to save people and protect property. The past and present firefighters have not

The Voices of IFD

In 1997 a quartet was formed by firefighter Dei Esther Passon to sing the national anthem at the Fire Department Instructors Conference (FDIC). Since establishing this quartet, voices of IFD, the original group made up of firefighters Daryl Hayden, Aleatha Henderson, Dei Esther Passon and Anthony Williamson have had the honor of representing the Indianapolis Fire Department and singing the national anthem at many events including the International Race of Champions (IROC), South Carolina Fireman's Convention, the Little League National Championship, 500 Mini Marathon, Governor's Conference, the Indiana Fire Chiefs Association Conference, Colts, Pacers, Fever, and Indians Games, and many Indianapolis Fire Department related functions.

1 January 2007

Former Washington Township Ladders were renumbered; #221 became #21, #223 became #6, and #224 became #4.

1 January 2007

The city agreed to a deferred compensation program for firefighters.

1 January 2007

The city agreed to print the Indianapolis Fire Department roster book by sonority every two years.

1 July 2007

Former Warren Township Engines were renumbered; #441 became #41, #442 became #42, #443 became #43, #444 became #44, and #445 became #45.

1 July 2007

Former Warren Township Ladder #443 is renumbered; became #43.

11 July 2007

Station #34 and Ladder #34 closed and reopens as Ladder #44.

2007

The Accountability Officer (AO) position was created.

FDIC Local 416 Open House and The Pumper Pull

The Local 416 Union opened its doors to the many firefighters that visit the city for the annual Fire Department Instructors Conference (FDIC) when the conference moved to Indianapolis in 1995. The first couple of years, the open house allowed firefighters to roam through the museum and buy a t-shirt or two while in town.

In 1998, the open house had their first pig roast for the firefighters. The following year in 1999, the open house included a pumper pull. This annual event has grown into one of the most anticipated days on the calendar for firefighters. The day-long event opens with drinks and food, followed by the pumper pull – a timed competition between teams that take turns pulling two antique fire trucks, and finishing with a live concert under the tent.

always made the correct decisions, but most would agree that becoming a firefighter with the IFD was one of their best. The citizens of Indianapolis should sleep well knowing that the people who watch over them and protect them in case of emergencies are proud, highly trained professionals, who want to be there.

The words Duty, Honor, Sacrifice, Brother, Pride, and Tradition aren't a part of the fire service because they sound inspiring, they are part of it because those are the words we live and might die by. They are words that we give meaning to. For the past 150 plus years all Indianapolis firefighters have lived it and they are linked together and for the next 150 years we hope that we have shaped the future IFD positively.

The tradition continues……

2008

A total of 289 car seats were installed by Survive Alive.

2008

Chief Brian L. Sanford became the 1st Chief from a merged department – Warren Township Fire Department.

2008

Department staffing consisted of 949 firefighters.

2008

Recruit Class #77 lasted 22 weeks.

2008

The Injured Firefighters Database was created.

2008

The Quartermaster Voucher Document System was implemented.

2008

The Shift Trainer Officer (STO) position was created.

2008

Training Academy was moved to 501 North Post Road.

2008

Unsafe Building Document System was established.

2 March 2009

Engine #26 closed.

2 March 2009

Squad #29 reopened, 3rd time.

1 August 2009

Former Perry Township Engines were renumbered; #661 became #61, #662 became #62, #663 became #63, and #664 became #64.

1 August 2009

Former Perry Township Ladders were renumbered; #663 became #63, and #664 became #64.

1 August 2009

Former Perry Township Squad #661 closed.

Cantaberry Apartment Fire

2009

Department staffing consisted of 1070 firefighters.

2009

Fire Prevention performed over 3622 total inspections.

2009

Staffing for one shift includes: 38 Stations, 37 Engines, 20 Ladders, 6 Squad's, 7 Battalion Chief's, 8 Medics, 4 Tactical Support Units (TSU), 3 Safety Cars, 2 Emergency Duty Officers (EDO), 1 Accountability Officer, and 1 Shift Commander.

2009

Station meals on average cost from $8.00 - $10.00 for both lunch and dinner.

2009

The 2nd Officer program was started.

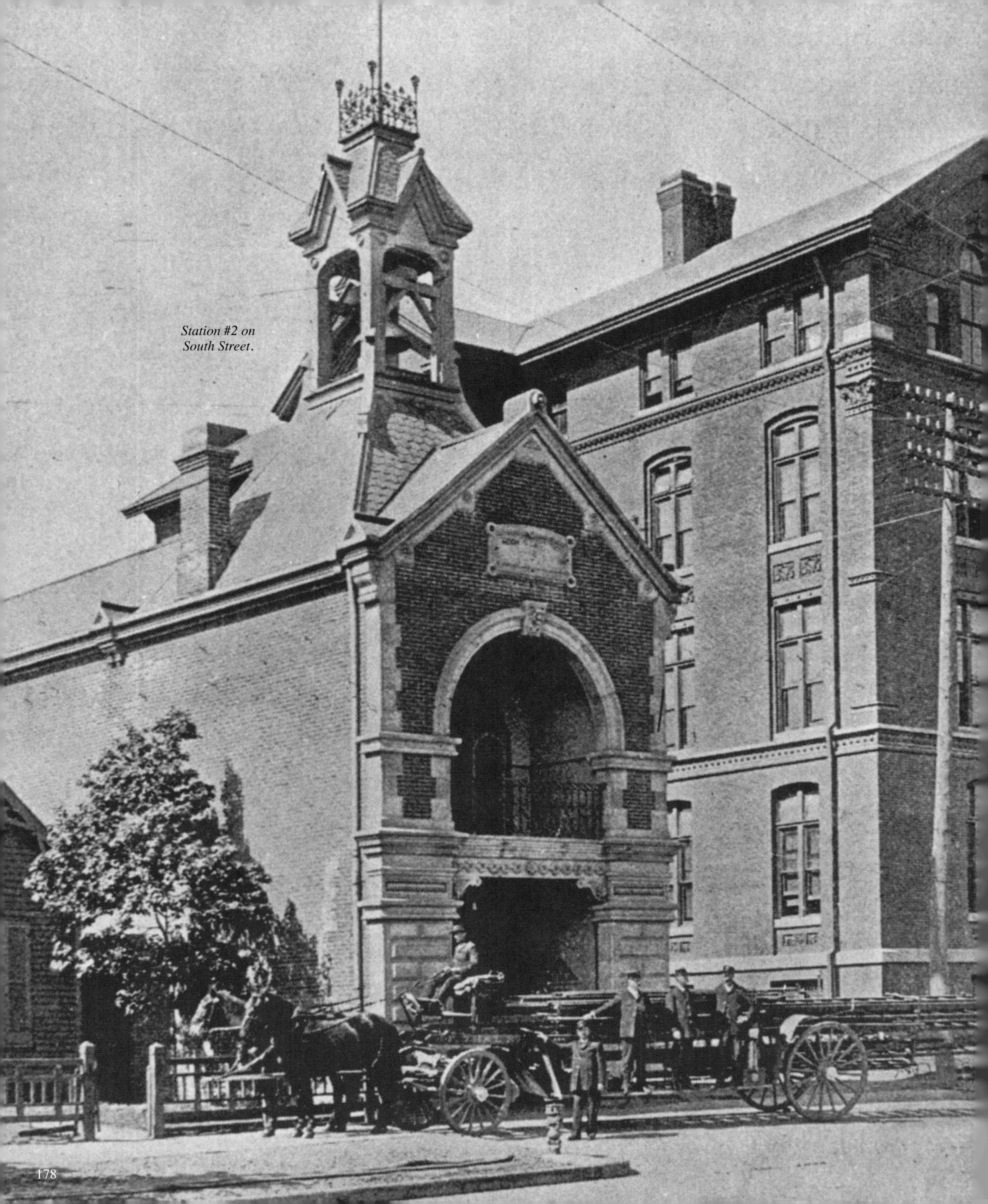

Station #2 on South Street.

The Firefighters' Quarters

The Marion Engine, Hose and Protection Company

The cities first fire station opened in 1835 on the north side of Monument circle just West of Meridian Street and housed the Marion's, a volunteer company. The station was burned to the ground in 1851, as a suspected arson, and was never rebuilt. The Marion's moved into a newly built station in 1855 on the Southwest corner of Massachusetts Avenue and North New York Street. This second Marion's station housed Company #2 in 1860 after the department was organized. In was later closed in 1872.

The Marion's

The Independent Relief Company

This station opened in 1841 on the West side of Meridian Street, South of Washington Street and housed the Independent Relief, a volunteer company. It was later closed in 1859.

The Hook and Ladder Company #1

This station opened in 1843 on Delaware Street between Market Street and Ohio Street and housed Truck Company #1, a volunteer company. It was later closed in 1860.

Former Station #9 and #16.

The Western Liberties

This station opened in 1849 on the point between Washington Street and The National Road and housed the Western Liberties, a volunteer company. It was later closed in 1857. The Western Liberties moved into a newly built station on the Southwest corner of Washington Street and West Street. This second Liberties station housed Company #2 1859-1860, Company #1 1860-1871, and Company #6 1871-1874. It was later was closed in 1874.

The O.K. Bucket Company

This station opened in 1849 on the Northeast corner of Meridian Street and Maryland Street and housed the O.K. Bucket Company, a volunteer company. It was later closed in 1859.

The Invincibles

This station opened in 1852 at 29 North New Jersey Street and housed the Invincibles, a volunteer company until the department was organized. It served as Station #1 from 1859-1860, Truck Company #1 1860-1872, and Hose Company #8 1872-1876 after the department was organized. It was later closed in 1876.

The Union's

This station opened in 1856 at 25 East South Street and housed the Union's, a volunteer company until the department was organized. It served as Company #3 1860-1872, and Company #7 1872-1874. It was later closed in 1874.

The Northwest Fire Company

This station opened in 1858 at Indiana Avenue on the canal and housed the Northwest Fire Company, a volunteer company. It was later closed in 1859. This station was never used by the paid department, but was closed as a result of the department being organized.

Reel House #9

This station opened in 1871 at 31 West St. Joseph Street and housed Hose Reel #9. It was the first fire station built by the paid department. It was later was closed in 1881. Company #9 moved to its newly built station at 1602 North Carrolton Avenue until it was renumbered Company #16 1897-1932, and temporary quarters for Engine #2 1935-1937. It was later closed in December of 1937.

Headquarters

In 1865, the Chief's office was located on the 2nd floor of the Glenblock building at 25 East Washington Street. The building at 250 Massachusetts Avenue was opened in the winter of 1872 and used as headquarters for the department. It was later closed in 1914. In 1914 Headquarters was moved to 301 East New York Street until June 1, 1979, when it moved to its new location at 555 North New Jersey Street.

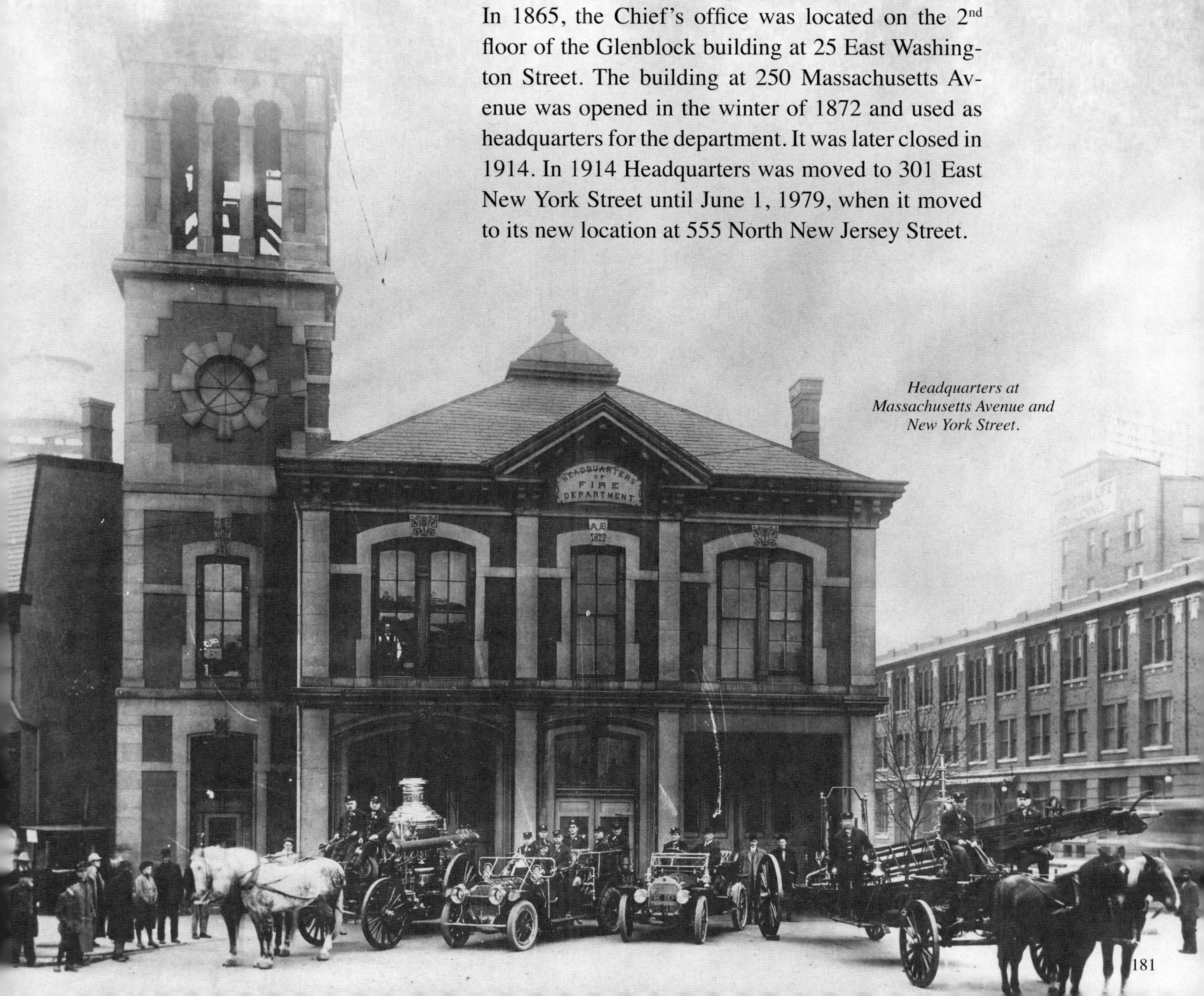

Headquarters at Massachusetts Avenue and New York Street.

Truck House #2

This station opened in 1874 at 125 East South Street and housed Truck Company #2. In 1907, it was renumbered station #28. It was later closed in June of 1910.

Reel House #8

This station opened 1876 at 533 East Washington Street and housed Hose Reel #8. It was later closed in 1896.

Reel House #10

This station opened in January 1881 at 1198 Madison Avenue and housed Hose Reel #10. It was renumbered Station #4 1890-1901, and Station #17 1901. It later closed on June 6, 1938.

Chemical Company #2

This station opened in July of 1883 at 1136 East Prospect Street and housed Chemical Company #2. In 1890 it was renumbered station #3. It was later closed on November 20, 1958.

The West Indianapolis City Hall and Fire Station

This station opened in 1894 at 1445 West Morris Street and housed the West Indianapolis Fire Company, a volunteer company. On April 1, 1897, the city of Indianapolis annexed this fire company and territory, making it part of the paid fire department. This station was renumbered station #19. It was later closed in May of 1931.

Chemical Company #4

This station opened in 1903 at 5430 East Washington Street and housed Chemical Company #4. The station was renumbered station #25 by years end. It was later closed in November of 1958.

Station #12 at
440 Virginia Avenue.

Station #1

The station at Washington Street and West Street opened as volunteer, Western Liberties Company in 1857. In 1860 it was renamed Station #1 until 1871 when it was renamed Station #6 until it closed in 1874. Station #1 moved to 441 Indiana Avenue on December 30, 1871, and remained open until October 20, 1938, when it was moved to 1445 West Michigan Street. Station #1 moved to its new location at 1903 West 10th Street on January 5, 1982. Station #1 is currently open in Battalion 13 and houses Engine 1, Ladder 1, and Tactical 1. Disciplines include Suppression, EMS, and Confined Space/Collapse and Trench Rescue (CRT).

Station #2

The station at Massachusetts Street and New York Street was opened as Volunteer, Marion's Fire Company #1 in 1855. In 1860 it was renumbered Station #2 and remained open until January 1, 1872, when it moved to 748 Massachusetts Avenue. On October 18, 1895, it was renumbered Station #8 until February 2, 1932, when it was closed. On October 18, 1895, Station #2 opened at 1531 North Roosevelt Avenue and remained open until August of 1935 when it was closed for a rebuild. On December 16, 1937, Station #2 reopened at 1575 North Roosevelt Avenue and remained open until March 22, 1971, when it moved to its new location at 4120 North Mitthoeffer Road. Station #2 is currently open in Battalion 4 and houses Engine 2. Disciplines include Suppression and EMS.

Station #3

The station at 25 East South Street opened as Volunteer, The Unions in April 1856. In 1860 it was renamed Station #3 and remained that until 1872 when it was renumbered Station #7 until it closed in 1874. On January 4th, 1872, Station #3 moved to 440 South Virginia Avenue until moving to 1136 East Prospect Street in 1890. This house, built on July 16, 1883, was previously occupied by Chemical #2. On November 20, 1958, Station #3 closed for a rebuild on the same property and reopened on October 20, 1959. Station #3 is currently open in Battalion 14 and houses Engine 3 and Battalion Chief 14. Disciplines include Suppression and EMS.

Station #4

"Pride of the North Side"

The station at 601 Russell Avenue opened as Hose Company #4 on January 4, 1872. It was renamed Station #10 in 1890 until it closed on October 16, 1941. In 1890, Hose Reel #10 located at 1198 South Madison Avenue was renamed Station #4 and remained until January 19, 1901 when it was renumbered Station #17 until closing on June 6, 1938. On January 19, 1901, Station #17 at 428 West Morris Street was renumbered to Station #4 until moving to 1761 South West Street on December 22, 1960. Station #4 was closed on March 31, 1992. On January 1, 2007, former Washington Township Station #224 at 8404 North Ditch Road became an Indianapolis Fire Department Station and was renumbered Station #4. Station #4 is currently open in Battalion 2 and houses Engine 4, Ladder 4, and Medic 4. Disciplines include Suppression, EMS, and Extrication.

Station #5

"16th Street Connection"

The station at 126 West 15th Street was opened as Station #5 on July 13, 1874. On December 5, 1933, Station #5 was closed to rebuild on the same property and reopened June 2, 1934. Station #5 moved to its new location at 155 West 16th Street on June 17, 1983. Station #5 is currently open in Battalion 13 and houses Engine 5 and Tactical 5. Disciplines include Suppression, EMS, and Confined Space/Collapse and Trench Rescue (CRT).

Station #6

"Protectors of the Crossing"

The station at Washington Street and West Street opened as Volunteer, Western Liberties in 1857. In 1860 it was renamed Station #1 until 1871 when it was renamed Station #6 and closed in 1874. On January 1, 1875, Station #6 moved to 533 West Washington Street until closing in March 1935. On October 15, 1952, Station #6 reopened at 933 West Washington Street. Station #6 was closed on December 11, 1974. On January 1, 2007, former Washington Township Station #223 at 1599 East 86th Street became an Indianapolis Fire Department Station and was renumbered Station #6. Station #6 is currently open in Battalion 2 and houses Engine 6, Ladder 6, Tactical 6, Boat 6 and Medic 6. Disciplines include Suppression, EMS, Extrication and Swift & Top Water Rescue.

Station #7

"Protectors of the Mile Square"

The station at 25 East South Street opened as Volunteer, The Unions in April 1856. It was renamed Station #3 in 1860 until being renumbered Station #7 in 1872 closing in 1874. In 1875 Station #7 moved to 26 East Maryland Street and was renumbered in 1896 to Station #13, returned to Station #7 from 1912 to 1915 before renumbering to Station #31 until closing in 1922. In 1914 Station #7 was at 301 East New York Street until June 1, 1979 when it moved to its new location at 555 North New Jersey Street. Station #7 is currently open in Battalion 14 and houses Engine 7, Ladder 7, Squad 7, Tactical 7, Boat 7, Accountability Officer - Car 1205, Safety 4 and the Shift Commander - Car 1201. Disciplines include Suppression, EMS, Heavy Extrication, High and Low Angle Rope Rescue, Swift & Top Water Rescue and Dive.

Station #8

"The Ocho"

The station at 29 North New Jersey Street was opened as Volunteer, The Invincibles in 1852. In 1860 in was renamed Truck House #1 until 1872 when it was renamed Hose Company #8. In 1895, Station #2 at 748 North Massachusetts Avenue, built in 1872, was renumber Station #8 and remained that until moving to 636 East 11th Street on February 27, 1932. Station #8 was closed on July 25, 1973. On January 1, 2007, former Washington Township Station #225 at 4045 East 56th Street became an Indianapolis Fire Department Station and was renumbered Station #8. Station #8 is currently open in Battalion 2 and houses Engine 8 and Safety 2. Disciplines include Suppression and EMS.

Station #9

The station at 31 East St. Joseph Street was opened as Station #9 on July 3, 1871. Station #9 moved to 1602 North Carrolton Avenue in January 1881 and remained until it was renumbered Station #16 on February 8, 1897, until closing April 21, 1932. In 1897, Station #9 moved to 1002 West Udell Street and remained until it was renumbered Station #23 on June 1, 1901, closing on January 10, 1961. On June 1, 1901, Station #20 at 537 North Belleview Place, built on March 1, 1898, was renumbered Station #9. Station #9 was permanently closed on January 5, 1982.

Station #10

"Brightwood Brigade"

The station at 1198 South Madison Avenue was opened as Reel House #10 on January 11, 1881. In 1890 it was renumbered Station #4 until January 19, 1901, when it was renumbered Station #17. It was closed on June 6, 1938. Station #4 at 601 Russell Avenue was renumbered Station #10 in 1890 and remained open until October 16, 1941 when it closed. Station #10 reopened at 3809 East 34th Street on August 25, 1958, and remained open until it moved to its new location at 2970 North Sherman Drive on February 7, 1997. Station #10 is currently open in Battalion 12 and houses Engine 10, Ladder 10, Squad 10, and Tactical Support Unit 10. Disciplines include Suppression, EMS, Extrication, and SQ10 only - Confined Space/Collapse and Trench Rescue (CRT).

Station #11

"Protectors' of Irish Hill"

The station at 440 South Virginia Avenue opened as Station #3 on January 4, 1872. In 1891 it was renumbered Station #11 and remained that until June 1, 1901, when it was renumbered Station #12 until closing on February 5, 1912. In 1901 Station #11 moved to 1030 East Washington Street which was built in 1896 and was home to Station #15. Station #11 moved to its new location at 1715 East Washington Street on April 27, 1971. Station #11 is currently open in Battalion 14 and houses Engine 11 and Car 1823 – EDO South. Disciplines include Suppression and EMS.

Station #12

The station at 352 North Beville Avenue opened as Station #12 in November 1896. In 1901 Station #12 moved to 440 South Virginia Avenue and remained until moving to 332 South New Jersey Street on February 5, 1912. In December 1915 this station was renumbered Station #30 and remained until closing April 21, 1954. Station #12 moved to 339 Sherman Drive in December 1915. Station #12 was closed on March 12, 1996. On January 1, 2007, former Washington Township Station #222 at 2151 Kessler Boulevard West Drive became an Indianapolis Fire Department Station and was renumbered Station #12. Station #12 is currently open in Battalion 2 and houses Engine 12 and Medic 12. Disciplines include Suppression and EMS.

Station #13

"Central Station"

The station at 26 East Maryland Street was opened as Station #7 in 1875. Station #7 was renumbered Station #13 in 1896. On April 13, 1912, Station #13 moved to 102 South Kentucky Avenue until being moved to 56 South Senate Avenue on December 24, 1969. Station #13 moved to its new location at 429 West Ohio Street on July 31, 1987. Station #13 is currently open in Battalion 13 and houses Engine 13, Ladder 13, Squad 13, Hazmat 13, and Battalion Chief 13. Disciplines include Suppression, EMS, and Hazmat

Station #14

"Protectors of Kenwood Forest"

The station at 2960 North Kenwood Avenue was opened as Station #14 on October 14, 1895. It was closed in April of 1971 for a rebuild, and reopened in February 7, 1972. It was again closed in 2000 and reopened in October 2001 at the same location making it the only station to be rebuilt three times on the same piece of property. Station #14 is currently open in Battalion 11 and houses Engine 14, Ladder 14, Squad 14, Tactical 14, and Boat 14. Disciplines include Suppression, EMS, Heavy Extrication, High and Low Angle Rope Rescue, Top Water Rescue and Dive.

Station #15

The station at 1030 East Washington Street was opened as Station #15 in 1896. In 1901 it was renumbered Station #11 and remained opened until April 27, 1971. In 1903 Station #15 moved to 2101 East English Avenue and remained until moving to its new location at 3502 East Prospect Street on April 28, 1971. Station #15 is currently open in Battalion 14 and houses Engine 15 and Ladder 15. Disciplines include Suppression, EMS, and Extrication.

Station #16

The station at 1602 North Carrolton Avenue was opened as Hose Company #9 in 1881. In 1897 it was renumbered Station #16 and remained opened until April 21, 1932 when it moved to its new location on the north side of town. From August 1935 until late 1936, this station became the temporary housing of the 2's. It was then closed permanently in December of 1937. On April 21, 1932, Station #16 moved to 5555 North Illinois Street. Station #16 is currently open in Battalion 2 and houses Engine 16 and Car 1822 - EDO North. Disciplines include Suppression and EMS.

Station #17

The station at 428 West Morris Street was opened as Station #17 on July 1, 1876. On January 19, 1901, it was renumbered Station #4 moving Station #17 to 1198 South Madison Avenue. On June 6, 1938, Station #17 closed until October 17, 1941, when it reopened at 1147 South Madison Avenue. Station #17 was closed permanently on February 17, 1984.

Station #18

©Chris Allen

The station at 1915 West Washington Street opened as Station #18 on July 1, 1896. On December 31, 1936, Station #18 was moved to 3130 West Washington Street until being moved to its new location at 42 North Warman Avenue on December 20, 1994. Station #18 is currently open in Battalion 13 and houses Engine 18 and Tactical Support Unit 18. Disciplines include Suppression and EMS.

Station #19

"Watchman of the Valley"

The station at 1445 West Morris Street was opened as the West Indianapolis Fire Town Hall and Fire Station in 1894. On April 1, 1897, it was renumbered Station #19 and remained until being moved to 1445 West Morris Street on December 24, 1931. Station #19 was moved to 1003 South White River Parkway West Drive on March 31, 1992. Station #19 is currently in open in Battalion 13 and houses Engine 19, Ladder 19, Decon 19 and Foam Truck 19. Disciplines include Suppression, EMS, and HAZMAT Decon.

Station #20

Task Force 20 "Let's Roll"

The station at 537 North Belleview Place was opened as Station #20 on March 1, 1898. On June 1, 1901, it was renumbered Station #9 until it closed on January 5, 1982. Old Station #12 at 352 North Beville Avenue was renumbered Station #20 in 1901 and remained open until it closed on October 17, 1942. Station #20 re-opened at its new location at 1452 North Emerson Avenue on September 18, 1952. Station #20 is currently open in Battalion 12 and houses Engine 20 and Ladder 20. Disciplines include Suppression, EMS, and Extrication.

Station #21

The station at 2360 North Olney Street was opened as Station #21 in March 1898 and remained opened until it moved to 3460 Glenn Drive on November 11, 1960. Station #21 was permanently closed on February 2, 1997. On January 1, 2007, former Washington Township Station #221 at 2508 East 71st Street became an Indianapolis Fire Department Station and was renumbered Station #21. Station #21 is currently open in Battalion 2 and houses Engine 21, Ladder 21, and Medic 21. Disciplines include Suppression, EMS, and Extrication.

Station #22

"Beast of the East"

The station at 2357 North Carrolton Avenue was opened as Station #22 on February 2, 1902. Station #22 was moved to its new location at 3019 North Martindale Avenue on August 31, 1972. Martindale Avenue was renamed Dr. Andrew J. Brown Avenue yielding a new address of 3019 North Dr. Andrew J. Brown Avenue. Station #22 is currently open in Battalion 11 and houses Engine 22 and Ladder 22. Disciplines include Suppression and EMS.

Station #23

"Pride of Riverside"

The station at 1002 West Udell Street was opened as Hose Company #9 on February 8, 1897. On June 1, 1901, it was renumbered Station #23 and remained open until January 10, 1961, when it moved to its new location at 975 West Burdsal Parkway. Station #23 is still currently open in Battalion 11 and houses Engine 23 and Battalion Chief 11. Disciplines include Suppression and EMS.

Station #24

The station at 1902 North Dexter Avenue was opened as Station #24 on March 11, 1903. Station #24 was moved to its new location at 5520 East 38th Street on October 27, 1961. Station #24 is currently open in Battalion 12 and houses Engine 24. Disciplines include Suppression and EMS.

Station #25

"Protectors of Irvington"

The station at 5430 East Washington Street was opened as Chemical #4 on September 1, 1903, and renamed Station #25 by years end. Station #25 was moved to its new location at 17 South Sheridan Avenue on November 9, 1958. Station #25 is currently open in Battalion 12 and houses Engine 25 and Battalion Chief 12. Disciplines include Suppression and EMS.

Station #26

The station at 2174 South Webb Street was opened as Station #26 in 1905. On April 23, 1964, Station #26 was moved to 1080 East Hanna Avenue. Station #26 was closed permanently on March 2, 2009.

Station #27

"Brookside's Bravest"

The station at 2918 East 10th Street was opened as Station #27 on January 1, 1907. On May 25, 1971, Station #27 closed for a rebuild on the same property and reopened on March 3, 1972. Station #27 is currently open in Battalion 14 and houses Engine 27 and Ladder 27. Disciplines include Suppression and EMS.

Station #28

The station at 125 East South Street was opened as Truck House #2 in 1876. In 1907 it was renumbered Station #28 and remained open until June 8, 1910, when it moved to its new location at 512 East Maple Road (38th Street). Station # 28 was closed permanently on March 19, 1995.

Station #29

"Garfield Park"

The station at 2302 South Shelby Street was opened as Station #29 on December 3, 1915. On April 17, 1991, Station #29 moved to its new location at 602 East Pleasant Run Parkway North Drive. Station #29 is currently open in Battalion 6 and houses Engine 29, Ladder 29, Squad 29, and Tactical Support Unit 29. Disciplines include Suppression, EMS, Extrication, and Urban Search and Rescue.

Station #30

"Wild Wild West"

The station at 332 South New Jersey Street was opened as Station #12 on February 5, 1912. In December 1915 it was renumbered Station #30 and remained open until April 21, 1954, when it was closed. Station #30 reopened on October 9, 1958, in its new location at 2440 North Tibbs Avenue. Station #30 is currently open in Battalion 11 and houses Engine 30 and Ladder 30. Disciplines include Suppression, EMS, and Extrication.

Station #31

"The Uptown Knights"

The station at 26 East Maryland Street was opened as Station #7 in 1875. Station #7 was renumbered Station #13 in 1896. In 1915 it was renumbered Station #31 until it closed on September 3, 1922. Station #31 reopened October 15, 1952, at 1201 East 46th Street and remained open until March 29, 1995, when it moved to its new location at 4155 North College Avenue. Station #31 is currently open in Battalion 11 and houses Engine 31, Ladder 31, Tactical Support Unit 31, and Hazmat 31. Disciplines include Suppression, EMS, and Hazmat.

Station #32

The station at 6330 North Guilford Avenue was acquired from the Broad Ripple City Hall and Fire Department on October 4, 1922, when it was annexed. Station #32 is currently open in Battalion 2 and houses Squad 32 and Battalion Chief 2. Disciplines include Suppression and EMS.

Station #33
"El Wild Wild West"

The station at 3430 North Moeller Road was opened on October 26, 1964. Station #33 is currently open in Battalion 11 and houses Engine 33. Disciplines include Suppression and EMS.

Station #34

The station at 3262 North Franklin Road was opened as Station #34 on April 26, 1967. Station #34 permanently closed on July 11, 2007.

Station #41

Former Warren Township Station #441 at 10750 East 10th Street was renumbered Station #41 on July 1, 2007. Station #41 is currently open in Battalion 4 and houses Engine 41 and Medic 41. Disciplines include Suppression and EMS.

Station #42

"We Bring our own Water"

Former Warren Township Station #442 at 1302 South Franklin Road was renumbered Station #42 on July 1, 2007. Station #42 is currently open in Battalion 12 and houses Engine 42, Tanker 42 and Medic 42. Disciplines include Suppression and EMS.

Station #43

"10th Street Express"

Former Warren Township Station #443 at 7604 East 10th Street was renumbered Station #43 on July 1, 2007. Station #43 is currently open in Battalion 4 and houses Engine 43, Ladder 43, Tactical 43, and Battalion Chief 4. Disciplines include Suppression and EMS.

Station #44

"Jokers Wild"

Former Warren Township Station #444 at 7345 East 30th Street was renumbered Station #44 on July 1, 2007. Station #44 is currently open in Battalion 4 and houses Engine 44, Ladder 44, Hazmat 44 and Medic 44. Disciplines include Suppression, EMS and Hazmat.

Station #45

"Our Family Protecting Your Family"

Former Warren Township Station # 445 at 10601East 30th Street was renumbered Station # 45 on July 1, 2007. Station # 45 is currently open in Battalion 4 and houses Engine 45 and Medic 45. Disciplines include Suppression and EMS.

Station #61

The station at 5410 South Madison Avenue was opened as Perry Township Station #1 in the early 1950's. Later renumbered Station #61, Station #61 moved to its new location at 1108 East Thompson Road in 1986. On August 1, 2009, Station #61 became an Indianapolis Fire Department Station. Station #61 is currently open in Battalion 6 and houses Engine 61, Battalion Chief 6, and Safety 6. Disciplines include Suppression and EMS.

Station #62

The station in the 7500 block of South Meridian Street was opened as Perry Township Station #2 in 1962. Later renumbered Station #62, Station #62 moved to its new location at 7447 South Meridian Street in 1990. On August 1, 2009, Station #62 became an Indianapolis Fire Department Station. Station #62 is currently open in Battalion 6 and houses Engine 62. Disciplines include Suppression and EMS.

Station #63

The station at 2707 East Stop 11 Road was opened as Perry Township Station #3 in 1968. Later renumbered Station #63, Station #63 moved to its new location at 4155 East Stop 11 Road in 1994. On August 1, 2009, Station #63 became an Indianapolis Fire Department Station. Station #63 is currently open in Battalion 6 and houses Engine 63 and Ladder 63. Disciplines include Suppression and EMS.

Station #64

The station at 1250 West Thompson Road was opened as Perry Township Station #4 in 1992. Later renumbered Station #64, Station #64 moved to its new location at 1925 West Edgewood Avenue in 2004. On August 1, 2009, Station #64 became an Indianapolis Fire Department Station. Station #64 is currently open in Battalion 6 and houses Engine 64, Ladder 64, and Tactical 64. Disciplines include Suppression, EMS, Extrication, and Top Water Rescue.

150th Year Celebration Sponsors

The Indianapolis Motor Speedway

The Indianapolis Motor Speedway Company was incorporated March 20, 1909, with the intent to build a track as a testing ground for Indiana's new automobile-manufacturing industry. The founders were four local businessmen: Carl G. Fisher, James A. Allison, Frank H. Wheeler and Arthur C. Newby. The 2.5-mile oval track was completed in mid-August 1909 with a tar and crushed-rock surface. Several months before the track was completed, however, the first race at IMS of any kind took place: a gas-filled balloon event on June 5, 1909. The first motorized race of any kind was a series of motorcycle races August 14, 1909; the first automobile races took place August 19, 1909.

The tar and crushed rock surface proved dangerous, however, thus track officials elected to have the track repaved in the fall of 1909 with 3.2 million 9.5-pound paving bricks. Before the job was completed, the track had already received its iconic nickname, the "Brickyard." Asphalt was applied to the track for the first time in spring 1936, and following A. J. Foyt's first Indianapolis 500 win in 1961, the entire track was repaved with asphalt except for the famed "Yard of Bricks" at the start-finish line. The original ownership group sold the track to Captain Eddie Rickenbacker on August 31, 1927, and the name of the company changed from IMS Company to IMS Corporation with the ownership change.

Following four years of inactivity during World War II (no races 1942-1945), Rickenbacker sold Indianapolis Motor Speedway Corporation to Anton "Tony" Hulman of Terre Haute, Indiana, on November 14, 1945. Despite more than four years of extreme neglect, Hulman's staff had the Speedway ready for resumption of the Indianapolis 500 in May 1946. Hulman, his wife, Mary Fendrich Hulman, their daughter Mari Hulman George, and her children Nancy George, Anton "Tony" George, M. Josephine George and Katherine George-Conforti, have continuously upgraded the Speedway's facilities to keep intact its status as "Racing Capital of the World."

Major change in the Speedway came when Tony George assumed presidency of IMS in 1990. The NASCAR Sprint Cup Series made its much-anticipated debut at IMS with the Brickyard 400 on August 6, 1994. IMS officials announced, December 2, 1998, plans to host the Formula One World Championship and also the largest facility improvement project in IMS history, which would include a new road course, Media Center, pit-side garages and suites, and other improvements. Formula One made its debut at IMS on September 24, 2000, with the United States Grand Prix at Indianapolis. The race would take place each year from 2000-2007. The MotoGP World Championship debuted at IMS on September 14, 2008, marking the first time motorcycles competed at the Speedway since August 1909.

As the Indianapolis Fire Department celebrates 150 years of operation, the Indianapolis Motor Speedway is in the midst of its Centennial Era, marking the 100th anniversary of the facility in 2009 and the 100th anniversary of the first Indianapolis 500 in 2011. The 100th running of the "500" will take place in 2016. The Speedway has for a century maintained a close and successful relationship with the Indianapolis Fire Department, as well as all of the city's public safety divisions. For all the men and women of IFD who have helped the Indianapolis Motor Speedway safely host millions of visitors and conduct events on a global stage, we say congratulations!

Kindred Healthcare and the IFD: A Shared Mission of Service

Kindred of Indianapolis Salutes the IFD for Service to the Community

Kindred Healthcare, Inc. is a healthcare services company based in Louisville, Kentucky with annual revenues of $6 billion and approximately 76,300 employees in 46 states. At June 30, 2011, Kindred through its subsidiaries provided healthcare services in over 2,200 locations, including 120 long-term acute care hospitals, five inpatient rehabilitation hospitals, 224 nursing and rehabilitation centers, 22 sub-acute units, 20 hospice and home care locations, 104 inpatient rehabilitation units (hospital-based) and a contract rehabilitation services business, RehabCare, which served approximately 1,760 non-affiliated facilities. Ranked as one of Fortune magazine's Most Admired Healthcare Companies for three years in a row, Kindred's mission is to promote healing, provide hope, preserve dignity and produce value for each patient, resident, family member, customer, employee and shareholder we serve.

Indianapolis Motor Speedway
(Photograph by John Cote)

Kindred has provided quality care and service excellence in the Indianapolis community for more than 15 years. Kindred understands that when people need healthcare services beyond their stay in a traditional hospital, they are often faced with a multitude of healthcare choices. To meet the individual needs of these often medically complex patients, we offer a diverse range of services

Our services include aggressive, medically complex care, short-term intensive rehabilitation and compassionate long-term care for dementia or Alzheimer's. Our nursing and rehabilitation centers offer long-term skilled care as well as short-term rehabilitation and skilled dementia care. All facilities have one goal – to help patients regain their highest level of independence in the shortest time period so that they can return to their previous living situation.

Kindred long-term acute care hospitals are specially designed for medically complex patients who require specialized care and extended recovery time. Our interdisciplinary care team consists of internists, rehabilitation therapists, social workers, dietitians and staff nurses.

Kindred Transitional Care and Rehabilitation – Bridgewater
14751 Carey Road
Carmel, Indiana 46033
317.575.2208
317.575.6165 fax
www.kindredbridgewater.com

Kindred Transitional Care and Rehabilitation – Castleton
5226 East 82nd Street
Indianapolis, Indiana 46250
317.842.6668
317.578.4113 fax
www.kindredcastleton.com

Kindred Transitional Care and Rehabilitation – Eagle Creek
4102 Shore Drive
Indianapolis, Indiana 46254
317.347.9051
317.347.9065 fax
www.kindredeaglecreek.com

Kindred Transitional Care and Rehabilitation – Greenfield
200 Green Meadows Drive
Greenfield, Indiana 46140
317.462.3311
317.462.8412 fax
www.kindredgreenfield.com

Kindred Transitional Care and Rehabilitation – Greenwood
377 Westridge Boulevard
Greenwood, Indiana 46142
317.888.4948
317.885.5085 fax
www.kindredgreenwood.com

Kindred Transitional Care and Rehabilitation – Wildwood
7301 East 16th Street
Indianapolis, Indiana 46219
317.353.1290
317.351.2579 fax
www.kindredwildwood.com

Kindred Hospital Indianapolis
1700 West 10th Street
Indianapolis, Indiana 46222
317.636.4400
317.636.4422 fax
www.kindredhospitalindy.com

Kindred Hospital Indianapolis South
607 South Greenwood Springs Drive
Greenwood, Indiana 46143
317.888.8155
317.888.7382 fax
www.khindysouth.com

As a company that seeks to serve the Indianapolis community, Kindred Healthcare appreciates and honors organizations with the same mission. So on September 11 Kindred of Indianapolis hosted a picnic luncheon honoring the Indianapolis Fire and Police Departments' heroic service to the Indianapolis community. The celebration, called "Kindred Salutes You Our Community Heroes!" was actually seven mini-events as each Kindred facility in the Indianapolis area held a luncheon to honor the IFD and IPD for helping to assure a high level of emergency care in Indianapolis. Guests, who included members of the Indianapolis Fire and Police Departments and their families, could choose which of the seven celebrations to attend for a cookout, door prizes and fun. And for those so busy serving the community they could not break for lunch, Kindred offered to-go boxes so they could continue doing what they do best. Kindred is proud to be a sponsor of the 150th Bereavement Ceremony honoring those that gave their lives in the line of duty. We salute our community heroes!

OneAmerica

For your tradition, protection and dedication... we offer our appreciation

Headquartered in Indianapolis for more than 130 years, the companies of OneAmerica continue to rely on the IFD for safety and security. Our 1200 employees thank you for providing a shield of protection that has helped make our OneAmerica Tower one of the safest high-rise buildings in the country.

We enthusiastically support the men and women of the Indianapolis Fire Department and celebrate your 150th anniversary. Your courage and sacrifice are an inspiration to those you serve.

As a financial services organization, we're focused on providing security, too. OneAmerica's nationwide network of companies offers a variety of products to serve the financial needs of their policyholders and other clients. These products include retirement plans, products and services; individual life insurance, annuities, long-term care solutions and employee benefits.

The products of the OneAmerica companies are distributed through a network of employees, agents, brokers and other distribution sources that are committed to increasing value to our policyholders by helping them plan to meet their financial goals.

We deliver on our promises when customers need us most.

OneAmerica salutes the heroes of the Indianapolis Fire Department.

150th Year Celebration Sponsors

150th Committee
92.3 WTTS
American Legion Post #64
Anglers World
Artistic Skin Design
Boogie Burger
Blue Ribbon Carriages
Cheeseburger in Paradise
Children's Museum
Coaches Tavern
Comedy Sport Complexions Spa Salon Boutique
Dicks Bodacious BBQ
Dicks Sporting Goods
Don Rabe
Dryer Honda
Embroidery Plus
Fast Times Indoor Karting
Firefighters Pest Control
FireRescue1.com
Firemen's Chore
Geist Reservoir
GG Food Services
Great Wolf Lodge
Greeks Pizza
Hoosier Fire Equipment
Hooters of Indianapolis
Holiday World Splashn Safari
IFD Emerald Society
IFD Fire Prevention
IFD Recruitment
IFD Station 3
IFD Station 5
IFD Station 7
IFD Station 29
Indianapolis Colts
Indianapolis Indians
Indianapolis Pacers
Indianapolis Pipes and Drums
Indy Acres
Indy Tactical Supply
Indy Victory/Triumph/BMW
International Wine Markets, Inc.
Keller Williams Reality
Kerasotes Showplace Theatres
Local 416
Local 669 - Sprinklerfitters
Maggiono's
Mass Avenue Pub
Metamorphisis Tattooing
Mild to Wild Pepper and Herb Company
North Central High School
Palominos
Penwell Publishing
Pike Township Fire Department
Police and Fire Insurance
Pro Image Sign and Design
Q'Doba Restaurant
Redline Apparel
Rolling Inferno MC Indy
Scotty's Brew House
Shane Hoskins
Shula's Steak House
Starbucks
Steve Lane IVFA
Sybaris
Sysco Foods
Tipton Sound and Lighting
Union Jacks Pizza
Walmart
Wayne Township Fire Department
Westside Bait and Tackle
WFMS
Wicked Motor Sports
Wonder Bread
Zeibart
Zore's Towing

Quotable Quotes

"Subie, you don't look old enough to be my paperboy."
"Subie, don't come back with your arms swinging."
"Where did you get your helmet painted?"
"Subie, I got underwear older then you."
"We don't even get that on a post card."
"Are you making a career out of it?"
"Get some time on the job."
"It's bustin with flavor."
"Never lay-out short."
"Subie, are you in?"
"Don't be a slug."
"Smoke Eater."
"Going under."
"Racked-out."
"Salty Dog."
"Doin it"

Photograph provided by Tod Parker (phototac.com)